INTRODUCTION TO THE STUDY OF MACROMOLECULES

INTRODUCTION TO THE STUDY OF MACROMOLECULES

Edited by

Dr. Kiran Abasaheb More

Delve PUBLISHING

www.delvepublishing.com

Introduction to the study of Macromolecules

Dr. Kiran Abasaheb More

Delve Publishing
224 Shoreacres Road
Burlington, ON L7L 2H2
Canada
www.delvepublishing.com
Email: orders@arclereducation.com

ISBN: 978-1-77469-863-1 (Hardcover)

ABOUT THE EDITOR

Dr. Kiran Abasaheb More, born in Bhogaon (Devi) Tahsil Jintur of Parbhani district of Maharashtra, India, in 1978, completed M.Sc in Botany, at Shree Shivaji College Parbhani, Maharashtra, Affiliated with Swami Ramanand Teerth Marathwada University Nanded, Maharashtra. The author also completed his Ph.D in 2010 in field of Taxonomy of Angiosperms and he joined the department of Botany, Yashwantrao Chavan Arts and Science Mahavidyalaya, Mangrulpir as an Assistant Professor. He is the Chief Editor of Research Journal Multilogic in science (Journal for Applied science), He edited more than 35 issues of research journals and authored 25 research papers. Dr. More is awarded by Young scientist award of Science and tech society for Integrated ruler improvement Telangana , and also awarded by Excellence in Teaching GENESIS ARBAN &RULER DEVOLOPMENT SOCIETY at the occasion of ICAAASTSD International conference 2018. He is the editor of proceeding of International conference on Advances in Agriculture and Allied science technologies for sustainable Development Society Special issue Jan 2018 and he is a Life Member of "GUARD Society". He also chaired a number of National and International Conferences, he is also a Member of Botany teacher association of Sant Gadge baba Amravati University Amravati, Maharashtra, India.

TABLE OF CONTENTS

LIST OF FIGURES

Figure 8.20. (Top) Schematic representations of the completely hydrosilated elastomeric network from 14 + 4 C–Ls (18) (left) and the partially hydrosilated elastomeric network from 17 + 4 C–Ls (19) (right). (Bottom) DSC thermograms depicting the completely hydrosilated networks from 14 + 4 C–Ls (18) (a), and the partially and completely hydrosilated elastomeric network (19) (b) and (20) (c), respectively, from 17 + 4 C–Ls reported by Kolel-Veetil et al.

Figure 8.21. Schematic representations of the uncomplexed (21a), partially complexed (21b), and completely complexed (21c) metallic derivatives of a diacetylene-containing carboranylenesiloxane.

Figure 8.22. The resistivity plot, the XRD spectrum and TEM micrographs of the pyrolysis of the $Cp_2\ Mo_2(CO)_4$ complex of 1.

LIST OF TABLES

LIST OF ABBREVIATIONS

3-D	Three-Dimensional
ADP	Adenosine-Diphosphate
AFM	Atomic Force Microscopy
AG	Arabic Gum
AMP	Adenosine Monophosphate
ATP	Adenosine Triphosphate
BSA	Bovine serum albumin
CMC	Carboxymethyl Cellulose
CMV	Cytomegalovirus
CWSP	Cell Wall Storage Polysaccharides
DDS	Drug Delivery Systems
DMS	Di-methyl-silane
DNA	Deoxyribonucleic Acid
DPPH	1,1-Diphenyl-2-Picrylhydrazol
EGF	Epidermal Growth Factor
EO	Essential Oil
FARG	Fragment Analysis Research Group
FGF	Fibroblast Growth Factor
GPC	Gel Permeation Chromatography
GPs	Glycoproteins
HMTS	Hexa-methyl-tri-siloxane
HSV	Herpes Simplex Virus
ISO	International Organization for Standardization
JRL	Jacalin-Related Lectins
LNP	Lectin Nucleotide Phosphohydrolase

MARG	Microarray Research Group
MDR	Multidrug-Resistant
MEM	Minimum Essential Medium
MIC	Minimum Inhibitory Concentration
MMP	Matrix Metalloproteinase
MSRB	Methionine Sulfoxide Reductase B
NARG	Nucleic Acid Research Group
NMP	N-Methyl-Pyrrolidone
NMR	Nuclear Magnetic Resonance
PAA	Poly Acrylic Acid
PSL	*Pisum sativum* Lectin
QC	Quality Control
RNA	Ribonucleic Acid
ROP	Ring-Opening Polymerization
RSV	Respiratory Syncytial Virus
RTV	Room-Temperature Vulcanization
SBL	Soybean Lectin
SCVP	Self-Condensing Vinyl Polymerization
SH	Sulfhydryl Groups
SNP	Single-Nucleotide Polymorphism
TEM	Transmission Electron Microscope
TG	Tragacanth Gum
TGA	Thermal Gravimetric Analysis
TMDS	Tetramethyl Di-siloxane
TNF	Tumor Necrosis Factor
UV	Ultraviolet
VAP	Viral Attachment Protein
VOCs	Volatile organic compounds
VZV	Varicella Zoster Virus

LIST OF GLOSSARY

A

Amantadine - A water-soluble crystalline substance, $C_{10}H_{17}NHCl$, used as an antiviral and antiparkinsonian drug.

Amide - It is an inorganic compound derived from ammonia by the replacement of an atom of hydrogen with another element (such as a metal).

Amorphous - In condensed matter physics and materials science, an amorphous or non-crystalline solid is a solid that lacks the long-range order, which is a characteristic of a crystal. In some older articles and books, the term was used synonymously with glass.

B

Bio-accessibility - Refers to the fraction of the total amount of a substance that is potentially available for absorption. In this study, laboratory methods were used to extract these bio-accessible portions from the soils. Bio-accessibility is therefore used to help predict bioavailability.

Bioavailability - Bioavailability is a subcategory of absorption and is the fraction of an administered drug that reaches the systemic circulation. By definition, when a medication is administered intravenously, its bioavailability is 100%.

Bio-membranes - A membrane either on the surface or interior of a cell that is composed of protein and lipid, especially in sheets only a few molecules thick and that limits the diffusion and transport of materials.

Buckyballs - A ball-like polyhedral carbon molecule of the type found in buckminsterfullerene and other fullerenes.

C

Cell Proliferation - Cell proliferation is the process by which a cell grows and divides to produce two daughter cells. Cell proliferation leads to an exponential increase in cell number and is, therefore, a rapid mechanism of tissue growth.

Chitin - Chitin is one of the most important biopolymers in nature. It is mainly produced by fungi, arthropods and nematodes. In insects, it functions as scaffold material, supporting the cuticles of the epidermis and trachea as well as the peritrophic matrices lining the gut epithelium.

Colloid - A colloid is a mixture in which one substance consisting of microscopically dispersed insoluble particles is suspended throughout another substance. Some definitions specify that the particles must be dispersed in a liquid, while others extend the definition to include substances like aerosols and gels.

Colloids - A colloid is a mixture in which one substance consisting of microscopically dispersed insoluble particles is suspended throughout another substance. Some definitions specify that the particles must be dispersed in a liquid, while others extend the definition to include substances like aerosols and gels.

Column Chromatography - In chemistry, Column chromatography is a technique that is used to separate a single chemical compound from a mixture dissolved in a fluid.

Copolymer - A copolymer is a polymer formed when two (or more) different types of monomers are linked in the same polymer chain, as opposed to a homopolymer where only one monomer is used.

Crystallization - Crystallization or crystallisation is the process by which a solid-forms, where the atoms or molecules are highly organized into a structure known as a crystal. Some of the ways by which crystals form are precipitating from a solution, freezing, or more rarely deposition directly from a gas.

Crystallographic - Crystallography is the experimental science of determining the arrangement of atoms in crystalline solids. Crystallography is a fundamental subject in the fields of materials science and solid-state physics.

D

Dimers - A dimer (di-, "two" + -mer, "parts") is an oligomer consisting of two monomers joined by bonds that can be strong or weak, covalent or intermolecular.

Dipoles - A pair of equal and opposite electric charges or magnetic poles of opposite signs separated especially by a small distance.

E

Elastomer - An elastomer is a polymer with viscoelasticity and with weak intermolecular forces, generally low Young's modulus and high failure strain compared with other materials.

Electronegativities - Electronegativity is a kind of measure that tells how strongly atoms attract the bonding electrons to them. Its symbol is a Greek letter that looks like an (X). It is directly proportional to the atom's attraction for the electrons.

F

Ferromagnets - It is the basic mechanism by which certain materials form permanent magnets, or are attracted to magnets. In physics, several different types of magnetism are distinguished

G

Gigantic – It refers to the size of a giant, or to size or scope befitting a giant.

Glycosidic - A glycosidic bond or glycosidic linkage is a type of covalent bond that joins a carbohydrate molecule to another group, which may or may not be another carbohydrate. A glycosidic bond is formed between the hemiacetal or hemiketal group of a saccharide and the hydroxyl group of some compound such as an alcohol.

H

Hydrolyze - Hydrolysis is any chemical reaction in which a molecule of water breaks one or more chemical bonds. The term is used broadly for substitution, elimination, and solvation reactions in which water is the nucleophile.

Hydro-silyation – Hydro-silylation, also called catalytic hydro-silation, describes the addition of Si-H bonds across unsaturated bonds. Ordinarily, the reaction is conducted catalytically and usually, the substrates are unsaturated organic compounds.

Hyper Physics – Hyper Physics is an educational website about physics topics. The information architecture of the website is based on HyperCard, the platform on which the material was originally developed, and a thesaurus organization, with thousands of controlled links and usual trees organizing topics from general to specific.

Hyperbranched macromolecules - Hyperbranched macromolecules (HMs, also called hyperbranched polymers) are highly branched three-dimensional (3D) structures in which all bonds converge to a focal point or core, and which have a multiplicity of reactive chain-ends.

I

Immobilization -The act of limiting movement or making incapable of movement.

K

Kinship - kinship is the web of social relationships that form an important part of the lives of all humans in all societies, although its exact meanings even within this discipline are often debated.

L

Leguminosae - A large family of dicotyledonous herbs, shrubs, and trees having fruits that are legumes or loments, bearing nodules on the roots that contain nitrogen-fixing bacteria, and including important food and forage plants (as peas, beans, or clovers)

Lignins - Lignin is an important organic polymer that is abundant in cell walls of some specific cells. It has many biological functions such as water transport, mechanical support and resistance to various stresses.

M

Malignancies - Malignancy is the tendency of a medical condition to become progressively worse. Malignancy is most familiar as a characterization of cancer.

Materials Science - Materials science is an interdisciplinary field concerned with the understanding and application of the properties of matter. Materials scientists study the connections between the underlying structure of a material, its properties, its processing methods and its performance in applications.

Metallocene - A metallocene is a compound typically consisting of two cyclopentadienyl anions bound to a metal center in the oxidation state II, with the resulting general formula $(C_5H_5)_2M$. Closely related to the metallocenes are the metallocene derivatives, e.g., titanocene dichloride, vanadocene dichloride

Monomers - A monomer is a molecule that can react together with other monomer molecules to form a larger polymer chain or three-dimensional network in a process called polymerization.

N

Nucleotide - Nucleotides are organic molecules consisting of a nuclEOside and a phosphate. They serve as monomeric units of the nucleic acid polymers – deoxyribonucleic acid and ribonucleic acid, both of which are essential biomolecules within all life forms on Earth.

O

Oscillation - Oscillation is the repetitive or periodic variation, typically in time, of some measure about a central value or between two or more different states. Familiar examples of oscillation include a swinging pendulum and alternating current.

P

Polymer - A polymer is any of a class of natural or synthetic substances composed of very large molecules, called macromolecules, which are multiples of simpler chemical units called monomers. Polymers make up many of the materials in living organisms and are the basis of many minerals and man-made materials.

Polymeric - A Polymeric substance or material consisting of very large molecules, or macromolecules, composed of many repeating subunits. Due to their broad spectrum of properties, both synthetic and natural polymers play essential and ubiquitous roles in everyday life.

Polypeptide - A polypeptide consisting of amino acids. Each polypeptide consists of a chain of amino acids linked together by covalent (peptide) bonds.

Proteases - A protease is an enzyme that catalyzes proteolysis, breaking down proteins into smaller polypeptides or single amino acids, and spurring the formation of new protein products. They do this by cleaving the peptide bonds within proteins by hydrolysis, a reaction where water breaks bonds.

R

Radicals - A free radical is an atom, molecule, or ion that has at least one unpaired valence electron. With some exceptions, these unpaired electrons make radicals highly chemically reactive. Many radicals spontaneously dimerize. Most organic radicals have short lifetimes.

Reagent - A reagent is an integral part of any chemical reaction. A reagent is a substance or compound that can facilitate a reaction, and they are used in most widely used tests. This includes, for example, pregnancy tests, blood glucose tests, and most COVID-19 test kits.

Rheological - Rheology is the study of the flow of matter, primarily in a fluid state, but also as "soft solids" or solids under conditions in which they respond with plastic flow rather than deforming elastically in reaction to an applied force.

Rimantadine - Rimantadine, a cyclic amine, is a synthetic antiviral drug and a derivate of adamantane, like a similar drug amantadine.Rimantadine is an orally administered antiviral drug used to treat, and in rare cases prevent, influenza virus A infection.

S

Single-Nucleotide Polymorphism - In genetics, a single-nucleotide polymorphism is a germline substitution of a single nucleotide at a specific position in the genome. Although certain definitions require the substitution to be present in a sufficiently large fraction of the population, many publications do not apply such a frequency threshold.

Spectroscopy - Spectroscopy is the general field of study that measures and interprets the electromagnetic spectra that result from the interaction between electromagnetic radiation and matter as a function of the wavelength or frequency of the radiation.

Stereoregularity - The degree to which successive configurations in space along the chain follow a simple rule.

T

Topological - Topology is a concept originally conceived in mathematics to describe unaltered spatial properties of an object under continuous deformation like bending, twisting, and stretching. Today, it has been used in many contexts with extensively stretched meanings.

V

Vasculatures - The blood vessels or arrangement of blood vessels in an organ or part.

Viscosity - Viscosity refers to the state or condition of being viscous. It is measured as a physical property of a fluid. As a physical property, it determines the internal resistance of the fluid to flow.

PREFACE

This book takes the readers through several aspects of Macromolecules. This book gives an introduction to select macromolecules, their structure and function, cisplatin derivatives as antiviral agents, synthetic versatility and structural modularity in organometallic polymers, plant macromolecules, functional applications of macromolecules, solid state macromolecules and advances in high temperature network polymers of Carboranylenesiloxanes and Silarylene Siloxades.

The first chapter stresses the basic introduction to macromolecules, so that the readers are clear about the philosophies behind them that form the utmost basics in the field. This chapter will also emphasize synthetic and biological polymers, macromolecular Science, macromolecular thermodynamics, natural macromolecules as carriers for essential oils, physical characteristics of EOs, bioavailability studies and needs for encapsulation of EOs.

The second chapter takes the readers through the concepts of Cisplatin derivatives as antiviral agents. This chapter will provide highlights on the various key aspects of inhibition, currently approved platinum-containing drugs, active form of cisplatin, structure activity relationships, arguments for polymeric drugs, cisplatin derivative drugs and polymer synthesis. The chapter also explains their antiviral activity, anticancer activity, spermicidal activity, fibers and experiments in biological characterization, synthesis and physical characterization.

Then, the third chapter explains macromolecules structure and function. It also explains the molecules of life, macromolecules are polymers, synthesis and breakdown of polymers, diversity of polymers, carbohydrates, lipids, proteins, nucleic acids and Sickle cell disease.

The fourth chapter introduces the readers to the synthetic versatility and structural modulation in organometallic polymers. This chapter also explains the polymerization, copolymerization of organometallic polymers, polymerization involving metal binding events during polymerization, synthetic pathways and polymeric organotin fibers. The chapter also sheds light on research, discussion, further considerations and outlook of the organometallic polymers.

The fifth chapter throws light on Plant macromolecules as biomaterials, plant polysaccharides, plant macromolecules as biomaterials for wound healing, plant-derived compounds, carbohydrates, proteins and recent advances using plant biomaterials for wound healing.

The sixth chapter takes the readers through the concept of functional applications of macromolecules. The readers are then told about chain length limitation and Up scalability aspects of discrete synthetic macromolecules, applications of discrete synthetic macromolecules in material science, self - assembly of discrete synthetic macromolecules in material science. The chapter also explains the foldamers based on uniform macromolecules, applications of discrete synthetic macromolecules in life science, pharmaceutical chemistry, macromolecular technologies, applications and improvements, antioxidative biomacromolecules and their applications.

The seventh chapter of the book explains solid state macromolecules, properties of solids, organization in the solid state, crystallinity, five types of crystalline solids, solid state of cross - linked macromolecules, the structure of configuration space for cross-linked system, construction of an order parameter, physical states and motions of small molecules and macromolecules.

In the last chapter of this book sheds lights on the advances in high temperature network polymers of Carboranylenesiloxanes and Silarylene Siloxades. This chapter also mentions various aspects of solid-state polymerization of Diacetyl groups, hydrosilylation reaction, carboranes, carboranylenesiloxane polymers containing thermally cross-linkable or vulcanizable diacetylene groups, hybrid siloxane network polymers from hydrosilylation reactions of siloxanes and carboranylenesiloxane monomers and its applications.

This book has been designed to suit the knowledge and pursuit of the researcher and scholars and to empower them with various aspects of select macromolecules, so that they are updated with the information. I hope that the readers find the book explanatory and insightful and that this book is referred by scholars across various fields.

CHAPTER 1

What Are Macromolecules

CONTENTS

Macromolecules and colloids have a strong relationship, and the two have been nearly inseparable in the past. Colloids were the first to be discovered, having been known for over a century. Macromolecules were only discovered after a long battle between chemists in the early 1900s.

While colloids and macromolecules are distinct entities, we now understand that many of the same principles that govern colloids also regulate macromolecules.

1.1. INTRODUCTION

Macromolecules have many of the same physical features as colloids, such as sedimentation, diffusion, and light scattering. Macromolecules have been thought of as linked colloids or lyophilic colloidal systems for many years. Macromolecules, on the other hand, are not colloids.

Colloids are clumps of tiny molecules that form as a result of a precise balance of weak attraction (van der Waals) and repulsive forces. Aggregation is influenced by the physical environment, especially the solvent. The aggregation may collapse if the solvent changes. Macromolecules are made up of several tiny molecules that are linked together by covalent bonds.

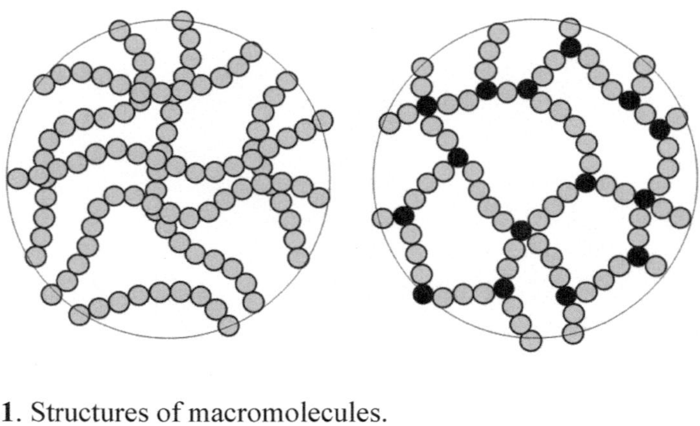

Figure 1.1. Structures of macromolecules.

Source: Image by Wikimedia Commons

Each macromolecule is a separate entity or unit, not a collection of them. A macromolecule's characteristics may differ as the solvent changes, but unless its covalent bonds are disrupted, the macromolecule remains a macromolecule. Synthetic polymers and biological polymers are the two most common forms of macromolecules.

Synthetic polymers are molecules created by humans and do not exist in nature. Although biological polymers can be found in nature, they can also be created in the lab. Synthetic polymers have only one or two similar repeating units in a chain, but biological polymers, especially proteins and enzymes, have many more identical repeating units in a chain (i.e., amino acids).

The molecules in synthetic polymers are usually not hard, and the chains are flexible. The polymer chains of biological polymers are more organized, and the molecules are stiffer in general. The rigidity of the chains is determined by their nature and their surroundings.

Nucleic acids are stiffer than proteins in comparison. Recently, further similarities between the two types of macromolecules have been discovered. Synthetic polymers, for example, that are typically thought to be in the shape of flexible random coils, can now be manufactured using Ziegler–Natta catalysts.

Synthetic polymers, which are typically thought of as flexible random coils, can now be synthesized with the Ziegler–Natta catalysts to achieve stereoregularity.

Furthermore, manmade polymers, like proteins and nucleic acids, can be created to have helices. The strong difference between synthetic polymers and biological polymers becomes increasingly arbitrary as our understanding of macromolecules grows.

1.2. SYNTHETIC POLYMERS

Carothers categorized synthetic polymers into two types in 1929, based on how they were made: condensation polymers and addition polymers. Condensation (or stepwise reaction) polymers are created by removing a minor molecule, such as water, from the reaction between two polyfunctional molecules. In a chain reaction of monomers with double bonds, addition (or chain reaction) polymers are generated. Polymers can be divided into two types based on their structural characteristics: linear polymers and branched polymers.

Commercially made synthetic polymers in the billions of pounds range can be divided into three categories:

1. Thermosetting resins (such as urea resins, polyesters, and epoxides) and thermoplastic resins (such as low-density and high-density polyethylene, polystyrene, and polypropylene);

2. Cellulosic (rayon and acetate) and noncellulose (polyester and nylon) synthetic fibers; and
3. Synthetic rubber (styrene–butadiene copolymer, ethylene–propylene copolymer, polybutadiene)

1.3. BIOLOGICAL POLYMERS

Amino acids, nucleotides, and sugars make up biological polymers. Proteins and polypeptides, nucleic acids, and sugar polymers are the three types of biological polymers discussed here.

1.3.1. Proteins and polypeptides

Amino acids are linked together by a peptide bond, which is an amide linkage between one molecule's amino group and another's carboxyl group. A polypeptide is a long chain of amino acids that makes up a protein. An enzyme is a protein that has catalytic properties.

Enzymes are proteins, but proteins are not all enzymes. Hormones are polypeptides (for example, insulin) that are closely related to proteins.

Simple proteins and conjugated proteins are the two types of proteins available. Simple proteins are classified into five classes based on their water solubility.

1. Albumins are water and dilute neutral salt solutions soluble proteins.
2. Globins (water-soluble proteins) (e.g., hemoglobin)
3. Globulins, which are water insoluble but soluble in dilute neutral salt solutions (e.g., g-globulins)
4. Prolamines, which are soluble in 70% ethyl alcohol but insoluble in water.
5. Histones—strongly basic, water-soluble solutions

Nonprotein groupings are used to characterize conjugated proteins:

1. Nucleoproteins—nucleic acid-binding proteins such as histones or Prolamines.
2. Phosphoproteins, which are proteins that are connected to phosphoric acid (e.g., casein in milk and vitellin in egg yolk)

3. Glycoproteins—a protein plus a carbohydrate [for example, mucin in saliva, mucoids in tendon and cartilage, and interferon, a human gene product synthesized in bacteria utilizing recombinant deoxyribonucleic acid (DNA) technology].
4. Chromoproteins—a protein that has a colorful component attached to it (e.g., Haemoglobin and cytochromes)
5. Lipoproteins are lipid-coated proteins (such as fatty acids, fat, and lecithin)
6. Membrane proteins—proteins that are incorporated in membranes' lipid core (e.g., glycophorin A)

Proteins come in three different shapes:

1. Narrow length (e.g., collagen, keratin, myosin, fibrinogen)
2. Spherical (e.g., serum albumin, myoglobin, lysozyme, carboxypeptidase, chymotrypsin)
3. Elasticity (e.g., elastin, the main constituent of ligament, aortic tissue, and the walls of blood vessels)

1.3.2. Nucleic Acids

Acids are made up of nucleotides, which in turn are made up of nucleosides. A base, a sugar, and a phosphate make up each nucleotide. A nucleotide can be made up of merely five bases, two sugars, and one phosphate. A nucleoside is a nucleotide that has been stripped of its phosphate.

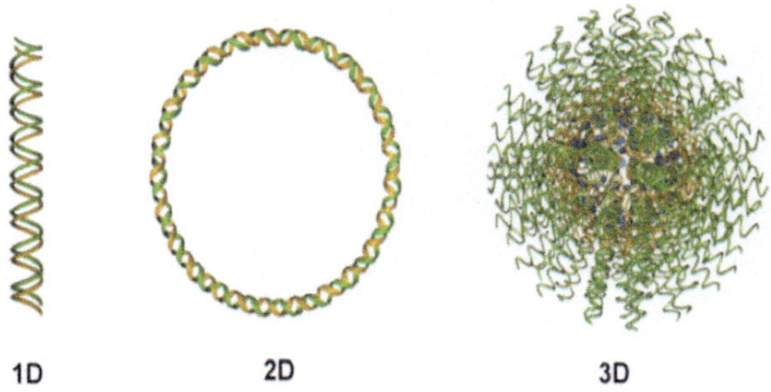

Figure 1.2. Classes of Nucleic Acids.

Source: Image by Wikimedia Commons

1.3.3. Polymers of Sugars

Polymers of sugars are frequently referred to as polysaccharides. They are monosaccharide polymers with a high molecular weight of 25,000–15,000,000.

The production of polysaccharides necessitates the production of hemiacetal and acetal. Hemiacetal is the product of an aldehyde reacting with an alcohol. A hemiacetal is transformed into an acetal after additional interaction with an alcohol.

The three homopolymers of glucose, starch, glycogen, and cellulose, are well-known polysaccharides. Amylose (produced by a-1,4-glucosidic linkage) and amylopectin (made by a-1,4-glucosidic linkage) are the two polymers that make up starch (a branched-chain polysaccharide formed by a-1,4-glucosidic bonds together with some a-1,6-glucosidic linkage).

Glycogen is a type of animal starch that is similar to amylopectin but has a more branching structure. Cellulose is a fibrous carbohydrate made up of D-glucose chains linked together by b-1,4-glucosidic bonds.

1.4. MACROMOLECULAR SCIENCE

Colloid science, surface science, and macromolecular science are the three fields of study that deal with colloids and macromolecules. Physical, mechanical, and chemical aspects of colloidal systems are studied in colloidal science. Surface science is concerned with macroscopic surface phenomena.

Macromolecular science studies the synthesis of synthetic polymers (or the isolation and purification of natural products such as proteins, nucleic acids, and carbohydrates) as well as the characterization of macromolecules. Polymer chemistry, polymer physics, biophysical chemistry, and molecular biology are only a few examples.

These three fields of study are intertwined. What one learns in one field is frequently applicable to others.

1.5. DISTRIBUTION OF MOLECULAR WEIGHT

There is no concern with molecular weight dispersion for tiny molecules like ethane. The molecular weight of all ethane molecules is the same. This is a difficulty with macromolecules, particularly synthetic polymers. Even if they are created in the same way, not all polymer molecules of the same substance will have the same molecular weight.

Any two molecular species (monomer and monomer, monomer and growing polymer, growing polymer and growing polymer, growing polymer and growing polymer) can react and cease at any time during step polymerization (condensation reaction).

As a result, multiple species of polymer molecules exist at the same time, ranging from those with extremely long chains to those with extremely short chains. A radical, an anionic, or a cationic reactive center adds a monomer unit to a developing polymer in a chain reaction (addition polymerization) as well.

Chain radical combination, chain radical disproportionation, and chain transfer to monomer, solvent, or any impurity are all examples of possible termination stages. These processes all result in polymer molecules with varying chain lengths.

While synthetic polymers have a severe difficulty with molecular weight distribution, proteins and nucleic acids do not. However, biological polymers in aqueous solutions frequently form dimers and trimers under particular conditions, implying that the solution is not homogeneous; for example, most bovine serum albumin (BSA) samples may contain 10% dimers.

The knowledge of molecular weight distribution could be used for biological polymers as well. The method of the distribution problem is statistical in nature, because molecular weight is directly proportional to chain length.

1.6. MACROMOLECULAR THERMODYNAMICS

The three most significant state functions to specify when characterizing the thermodynamic behavior of a system (here, macromolecules) are S, H, and G. The partial molar amounts of these three functions, especially G (the chemical potential), must be added as the grounds for the understanding of polymer phenomena when applied to solutions.

We'll start with a general overview of thermodynamic principles in this chapter. This is essentially a refresher course on thermodynamic terminology. We describe Flory's lattice theory of S of mixing, his idea of contact energy H of mixing, and his use of G and G in the handling of dilute polymer solutions in detail because he, among others, created the cornerstone of physical polymer chemistry. For many years to come, his contribution to the formation and explanation of the two parameters w1 and y has left an indelible mark on polymer language. Before commencing on any advanced subject, we believe it is beneficial for a reader or researcher to get familiar with Flory's work.

We discuss Hildebrand's theory of the solubility parameter d, not just for comparisons but also for applications, in conjunction with Flory's concept of contact energy.

1.6.1. Review of Thermodynamics

Let's start by defining two key words in thermodynamics: system and environment. A system or thermodynamic system is any material body under investigation (e.g., 1 mol of gas, 500 mL of protein solution). The rest of the world is referred to as the surroundings in relation to a certain system. The cosmos is made up of a system and its surroundings.

Thermodynamics is the study of how energy changes in a system as it interacts with its environment. To characterize a system, we need three parameters: P (pressure), V (volume), and T (temperature) (temperature). A fourth metric, C (heat capacity), is chosen for convenience and is closely connected to T.

We may now construct two thermodynamic quantities, w (work) and q (heat), in differential equation form, using P, V, and T as three independent variables and C as an auxiliary variable.

Dw = P dV **dq = C dT**

The symbol w stands for work performed on or by the system, and the letter q stands for heat (energy) entering or exiting the system. W or q is always tied to the surroundings, whether it's 'done on or by' or 'entered or freed from.'

Positive and negative signs can be seen in the quantities w and q. When heat from the surroundings enters the system, q is positive; when heat is discharged from the surroundings, q is negative. Similarly, if the system works on the surroundings, w is positive; if the surroundings operate on the system, w is negative.

If the three independent variables are specified, we say the state of a system is known. When we know the exact values of P1, V1, and T1, for example, 1 mol of gas is in state 1. Similarly, if we know the exact values of P2, V2, and T2, the same 1 mol of gas is in state 2. When the values of P1, V1, and T1 change to P2, V2, and T2, the system changes. We call this transition from state 1 to state 2 of the system. The term path is used in calculus to refer to the path used to get from state 1 to state 2. Because the two values dw and dq both depend on the path, this term is crucial to our discussion. To transition from state 1 to state 2, a system can take a variety of paths.

As a result, assuming all other conditions are identical, path 1 from state 1 takes significantly less work (dw) and energy (dq) than path 2 or path 3. While both dw and dq are reliant on the path, the combination of the two values in the form of dq makes dw independent of the path. The dq, dw combination is solely determined by the states involved.

If a change happens from state 1 to state 1 by a specific path, dq and dw are not zero, but dq dw is zero because no change in state occurs. We employ a specific term, dE, to express dq dw because of its importance:

$$\mathbf{dE = dq - dw}$$

The first law of thermodynamics is this. The letter E stands for internal energy.

All processes that change a system's state from one to another can be divided into two categories: reversible and irreversible. There are an endless number of equilibrium states in a reversible process from state 1 to state 2, but not in an irreversible process.

In other words, in a reversible process, a system can return to any previous equilibrium condition, whereas in an irreversible process, the energy shift cannot be reversed.

1.6.2. Partial Specific Volume

In terms of molecular structure, there is no perfect description of partial specific volume. The parameter vi appears to be insufficient to characterize the behavior of component I in the solution in terms of size, shape, and chemical reactivity. However, as we will see, it is an essential variable that allows us to interpret other experimental factors such as light scattering, diffusion, and sedimentation.

However, as we will see later, it is a necessary value for interpreting other experimental factors such as light scattering, diffusion, and sedimentation. Density gradient columns, pyconometry, magnetic flotation methods, and vibration methods are some of the experimental methods for determining partial specific volume. The measurement process is typically laborious and challenging, regardless of which approach we choose.

1.7. NATURAL MACROMOLECULES AS CARRIERS FOR ESSENTIAL OILS: FROM EXTRACTION TO BIOMEDICAL APPLICATION

Plant-based goods are used by an estimated 80% of the world's population in official and traditional medicine, with plant-based medicines accounting for around a quarter of the total pharmaceutical arsenal (Bhattaram et al., 2002). Plant bioactive molecules are also widely used in the pharmaceutical, cosmetics, and food industries, as well as fine (agro) chemicals and nutraceuticals (Bourgaud et al., 2001).

Isoprenoids are the biggest class of plant bioactive chemicals, and they have a role in the composition of essential oils (EOs) in a variety of ways. Because of certain of its features, including volatility, pungent odor and taste, dose-dependent toxicity, and very high biological activity, encapsulation of EO is a must for using EOs.

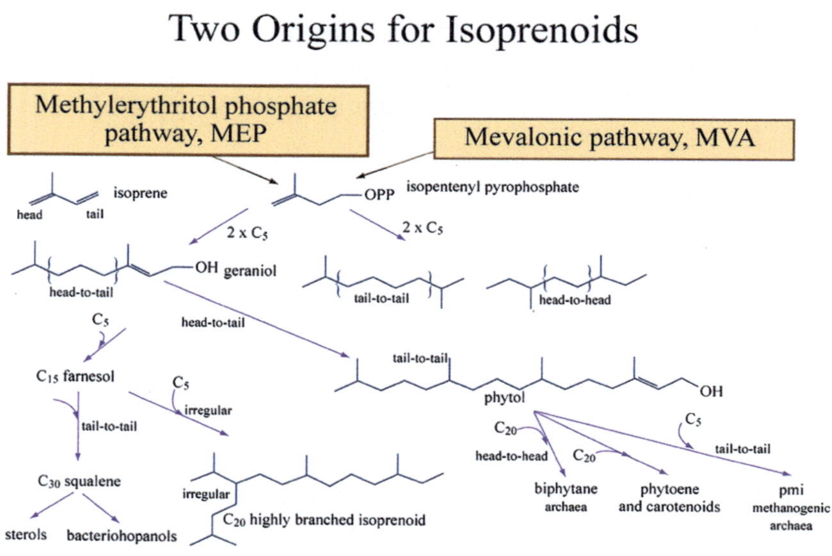

Figure 1.3.Two Origins for Isoprenoids.

Source: Image by Flickr

The encapsulation also shields the body from the negative effects of EOs by preventing its breakdown and transformation during digestion (Maderuelo et al., 2019). Entrapping EOs also improves their bioavailability.

The development of numerous encapsulation techniques assures that EOs is widely used in the pharmaceutical and cosmetics industries (e.g., Asbahani et al., 2015; Pandit et al., 2016; Arpagaus et al., 2018). (e.g., Martins et al., 2014; Carvalho et al., 2016). The overall engineering optimization of the delivery process consists of multiple interconnected steps:

1. Selection of carrier matrix with the highest performance under specific delivery conditions
2. Ensuring the highest level of encapsulation efficiency possible, and
3. Reusability of carrier matrix components such as natural polysaccharides and proteins is taken into account.

Polysaccharides like pectin, inulin, starch, cellulose, and hemicelluloses are commonly employed in single hydrogels and blends. Various chemical or enzymatic approaches can be used to remove these macromolecules from wastes from vegetable industrial processing (Poli et al., 2011).

Polysaccharide hydrogels have been combined with natural proteins such as soy proteins, whey proteins, lecithin, and others to increase their chemical stability under the gastric surroundings for medicinal and biotechnological applications like EO encapsulation.

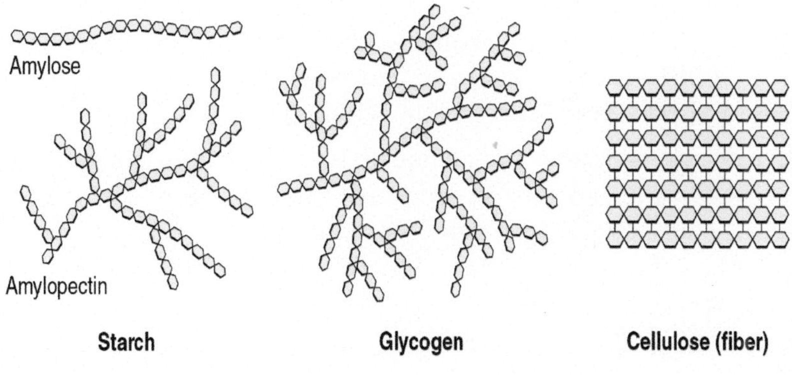

Figure 1.4. Three Important Polysaccharides.

Source: Image by Wikimedia Commons

EOs are products derived from vegetable raw materials through physical processes like distillation or pressing, according to the International Organization for Standardization (ISO). According to the EssOilDB database, plant volatiles are characteristic products of aromatic plants that have been identified in 1,618 plant species, subspecies, or variations representing 92 plant families (Kumari et al., 2014).

Essential oils are complex mixes made up primarily of volatile organic compounds (VOCs) produced and exhaled by plants to aid their growth and survival (Loreto et al., 2014). Aside from VOC, EOs contains a variety of degradation products produced by enzymatic, chemical, and physical activities.

The breakdown of matricine during steam distillation produces chamazulene, which is a good example (Clarke, 2008). The other example is khusimone, nor-patchoulenol, or nor-tetrapatchoulol, all of which are sesquiterpenoids with only 14 carbon atoms (Baser and Buchbauer, 2010).

Essential oils are combinations of low-molecular-weight compounds (typically less than 300 Da) made up of dozens, if not hundreds, of molecules belonging to five to ten different chemical classes or congeneric groupings.

In the EOs, the content of a single constituent or a congeneric collection of constituents can range from tenths of a percent to several dozen percent (Baser and Buchbauer, 2010; Dhifi et al., 2017).

Isoprenoids, phenylpropanoids, polyketides, and lipids are among the constituents of EOs, which come from three plant biochemical processes (Dudareva et al., 2006; Baser and Buchbauer, 2010; Moghaddam and Mehdizadeh, 2017). L-phenylalanine converts glucose, which is generated from carbon dioxide and water, into phosphoenolpyruvate, involved in the synthesis of phenylpropanoids (shikimates).

After decarboxylation, the identical structure of phosphoenolpyruvate yields acetate, which esterifies with coenzyme-A to yield acetyl-CoA. The self-condensation of acetyl-CoA produces polyketides and lipids. Isoprenoids are formed from acetyl-CoA, which is also utilized to make mevalonic acid (Baser and Buchbauer, 2010).

Apart from L-phenylalanine, the other components of EOs are amino acid derivatives (Dudareva et al., 2006).

1.7.1. Isoprenoids

Isoprenoids (terpenes) are made up of isoprene (2-methylbutadiene) units joined together in a pattern known as head-to-tail joining, and their structure has a multiple of five carbon atoms (Mann et al., 1994; Baser and Buchbauer, 2010). As a result, the amount of isoprene units in a molecule determines the structural classification of terpenes.

Monoterpenes have 10 carbons (C10), sesquiterpenes have 15, and diterpenes have 20. Hemiterpenes are made up of one isoprene unit (C5); monoterpenes have 10 carbons (C10), sesquiterpenes have 15 (C15), and diterpenes have 20 (C20).

The type of substituents, the degree of unsaturation, and the cyclic or linear structure all influence functional characterization (hydrocarbons, alcohols, ethers, oxides, aldehydes, ketones, esters). Isoprenoids with an oxygen moiety are referred to as terpenoids (Baser and Buchbauer, 2010; Moghaddam and Mehdizadeh, 2017).

Some of the molecules in EOs are decomposition products of bigger, typically non-volatile compounds. Norisoprenoids are breakdown products of triterpenoids or tetraterpenoids that occur as a result of enzymatic or non-enzymatic cleavage.

Irones are the breakdown products of the triterpenoid iripallidal, whereas ionones, damascones, or megastigmanes are the degradation products of the center section of the carotenoid chain.

1.7.2. Phenylpropanoids

The C6–C3 benzene ring unit (C6) is connected to a three-carbon side chain (C3) at position 1 and oxygenated at the third/fourth/fifth position/s in shikimate derivatives. A carbon–carbon double bond is common in C3, but the side chain can be reduced to just one carbon (C1).

Phenols, also known as phenol ethers, are phenylpropanoids that are commonly found in EOs (Baser and Buchbauer, 2010; Moghaddam and Mehdizadeh, 2017).

1.7.3. Derivatives of Polyketides and Lipids

Condensation processes of polyketides, lipid breakdown, and arachidonic acid cyclization produce fatty acid derivatives in EO (Dudareva et al., 2006). When polyketides condense, phenolic rings develop, which are oxidized on alternate carbon atoms to produce acids, ketones, phenols, or one end of a double bond (Baser and Buchbauer, 2010).

Short-chain lactones, alcohols, and aldehydes are produced by a variety of enzymatic processes on fatty acids, including cleavage, oxidation, lactonization, reduction, and elimination, whereas prostaglandins and jasmonates are produced by cyclization of arachidonic acid (Dudareva et al., 2006; Baser and Buchbauer, 2010)

1.7.4. Derivatives of Amino Acids Other Than L-Phenylalanine

Aldehydes, alcohols, esters, acids, and nitrogen- and sulfur-containing elements of EOs are all formed from amino acids such as alanine, valine, leucine, isoleucine, and methionine (Dudareva et al., 2006).

Sulfur compounds (sulfide, disulfides, trisulfides, sulfoxides, and isothiocyanates) are rare in EOs, as are heterocyclic compounds with nitrogen (indole, methyl anthranilates, pyridines, and pyrazines) or oxygen (lactones, coumarins, and furanoids) in a ring. These compounds have simple structures and a strong distinctive or pungent odor (Moghaddam and Mehdizadeh, 2017).

1.8. PHYSICAL CHARACTERISTICS OF EOS

The odor of EOs is their most distinguishing feature. Due to the high vapor pressure at atmospheric pressure and room temperature, EOs constituents are partially in the vapor state (Dhifi et al., 2017).

Volatility is reduced as the number of carbon atoms in the structure is increased. Monoterpenes have the greatest boiling temperatures, making their molecules extremely volatile and evaporating quickly. Sesquiterpenes are still volatile enough to be present as EOs, although diterenes are less common in volatile fractions (Baser and Buchbauer, 2010).

Plant material is normally distilled into a clear, colorless, or pale-yellow liquid that is immiscible with water and has a lower density than water. There are, however, certain exceptions. Orris and guaiac wood, as well as plumeria, are used to make solid or semisolid EOs.

Cinnamon produces a yellow to brownish EO, whereas chamomile produces a blue EO, European valerian produces a green EO, and vetiver produces a brown EO. Cinnamon, sassafras, and vetiver EOs have a density of one or close to one when compared to water (Dhifi et al., 2017; Moghaddam and Mehdizadeh, 2017). Fats, alcohols, and most organic solvents are all soluble in essential oils. Asymmetric carbons are present in their constituents, resulting in optical activity (optical rotation). They are also distinguished by their refractive index (Moghaddam and Mehdizadeh, 2017). The characteristics used to control EOs quality include density, optical rotation, and refractive index. The range of allowable levels for the most widely available EOs is described in ISO1 standards (ISO/TC 54: ISO 279:1998; ISO 280:1998; ISO 592:1998).

1.8.1. Stability of EOs

Volatiles are vulnerable to temperature, light, oxidation, and hydrolysis when they are released from plant structures (ducts or glands). The chemical content of plant material, as well as plant material processing and storage, distillation methods, and subsequent EOs handling, all influence the final composition of EOs (Turek and Stintzing, 2013; Moghaddam and Mehdizadeh, 2017).

The chemical composition of EOs components is the most important element influencing their stability. Because hydrogen atom abstraction leads in resonance-stabilized radicals, compounds with double bonds are prone to autoxidation. Conjugated double-bonds can stabilize radicals formed by polyunsaturated terpenic hydrocarbons.

Simultaneously, isomerization to tertiary radicals might occur, resulting in oxidative degradation (Turek and Stintzing, 2013). Access to atmospheric oxygen triggers spontaneous free radical chain reactions, resulting in unstable hydroperoxides that disintegrate in the presence of light, heat, or rising acidity.

Secondary oxidation products that are stable include monovalent to polyvalent alcohols, aldehydes, ketones, epoxides, peroxides, acids, and oxygen-bearing polymers. Some oxygen-bearing terpenoids, on the other hand, are transformed straight into oxidized secondary products without the formation of hydroperoxides (Geier, 2006; Turek and Stintzing, 2013).

Because headspace oxygen diffuses into the sample over time, the EOs should be maintained in totally filled containers or, if possible, treated with inert gas to remove any leftover air and prevent oxidative reactions (Geier, 2006; Turek et al., 2012). Light and temperature are the other two elements that are strongly linked to EOs oxidative degradation.

Light enhances autoxidation and the generation of alkyl radicals in monoterpenes, catalyzes intramolecular isomerization events or trans–cis conversions, and boosts monoterpene degradation (Turek and Stintzing, 2013).

Heat speeds up chemical reactions and aids in the development of primary auto-oxidation products, such as hydroperoxides, which are then degraded when the temperature rises, yielding final oxidation products (Turek and Stintzing, 2012, 2013; Turek et al., 2012). At high temperatures, volatiles are thermolabile and vulnerable to rearrangement processes. Cleavage of double bonds, epoxidation, dehydrogenation into aromatic systems, and allylic oxidation into alcohols, ketones, and aldehydes are the four types of oxidative processes that terpenes go through during thermal deterioration (McGraw et al., 1999).

The production of alkyl or hydroxyl radicals is more apparent at higher temperatures because oxygen solubility is lower. On the other side, storing EOs at low temperatures promotes oxygen solubility in liquids, resulting in the formation of peroxide (Turek and Stintzing, 2013).

Compounds, primarily isoprenoids, easily oxidize in complex mixes like EOs, affecting more stable structures and causing rearrangement and breakdown events. In exchange, the phenylpropanoids in EOs work as antioxidants, scavenging free radicals and protecting other molecules from degradation (Turek and Stintzing, 2013).

EOs are deteriorating in quality as a result of the above-mentioned breakdown processes. Changes in color, consistency, and odor are the most visible indications of age, with the latter being particularly unpleasant and strong.

The biological activity of EOs is significantly influenced by its general physicochemical characteristics (complexity and interactions of individual compounds) and ingredients (low molecular weight, presence of diverse functional groups in the molecule, reactivity, and hydrophobicity).

1.8.2. Bioavailability of EOs

Essential oils and/or their already extracted individual constituents are predominantly studied for their antibacterial and antioxidant action (e.g., Burt, 2004; Nazzaro et al., 2013; Semeniuc et al., 2017). (e.g., Miguel, 2010).

EOs' antimicrobial capabilities are linked to their capacity to penetrate the bacterial cell wall, where rupture of the bacterial cell wall causes ion leakage, membrane potential lowering, disruption of membrane enzymes, and structural and functional changes in the bacterial cell (Edris, 2007).

The antioxidant capabilities of EO compounds are linked to their H-donating properties, ability to suppress lipid autoxidation, singlet oxygen quenching, hydrogen transfer, or electron transfer (e.g., Grassmann, 2005).

Apart from these effects, EOs have been extensively researched for their anticancer (e.g., Bayala et al., 2014), anti-inflammatory (Sá et al., 2014), anxiolytic (De Sousa, 2012), analgesic-like (De Sousa, 2011), antinociceptive (Lenardo et al., 2016), antiaging, and neuroprotective properties (Ayaz et al.,

2017). Several comprehensive papers described the complex and diverse bioactive roles of EOs and their products (e.g., Edris, 2007; Bakkali et al., 2008; Raut and Karuppayil, 2014; Sarkic and Stappen, 2018). However, little is known about the activity of EOs and their constituent compounds in the human body after ingestion.

Bioavailability relates to the drug's pharmacokinetic qualities when it enters the systemic circulation, allowing it to act at the target sites (Stahl et al., 2002). In fact, bioavailability consists of two interrelated subsets: bio accessibility and bioactivity. The quantity or fraction of food matrix released in the gastrointestinal tract that is available for absorption is referred to as bio accessibility (Thakur et al., 2020). It also represents the compound's potential for assimilation following digestive transformations, absorption, and presystemic intestinal and hepatic metabolism (Cardoso et al., 2015).

The procedure of a drug entering systemic circulation, transportation to the target site, and interactions with diverse biomolecules that result in the manifestation of various metabolic and physiological effects is referred to as bioactivity (Wood, 2005; Carbonell-Capella et al., 2014).

Bioavailability is the most important factor to consider when evaluating medication absorption through various modes of delivery (Maderuelo et al., 2019).

1.9. APPROACHES IN BIOAVAILABILITY STUDIES

Different *in vivo* and *in vitro* approaches are used to investigate the bioavailability and bio-accessibility of plant metabolites, including EOs and their constituent terpene components. Bio-accessibility is frequently estimated using *in vitro* digestive models.

The majority of these methods simulate GI system conditions by adjusting pH and introducing specific digestive enzymes (e.g., salivary amylase, pepsin, gastric lipase, trypsin, chymotrypsin, pancreatic lipase, etc.), bile salts, and, in some cases, fermentation reactions to replicate colon performance (Jones et al., 2019).

Cell models, principally Caco-2 cells derived from human colorectal adenocarcinoma, have been used in recent *in vitro* bio-accessibility/ bioavailability investigations, in which the absorbed target drug is collected on the basolateral side of monolayer model cells (Jones et al., 2019; Thakur et al., 2020) *In vivo* animal and clinical research have also been used to investigate bioavailability. Bioavailability of EOs and its compounds, like that of other drugs, refers to the concentration threshold that reaches the blood circulation system and includes digestion (in the case of oral administration), absorption, metabolic transformation, tissue distribution, and bioactive performance at the target sites (Carbonell-Capella et al., 2014).

The pharmacodynamic and pharmacokinetic properties of bioactive substances are included in bioavailability. EOs pharmacodynamics involves the effects of specific chemicals on human and animal biochemical and physiological processes, i.e., biological activity monitoring at target organs, tissues, and cells.

Regardless of EOs' limited selectivity, their bioactive effects are manifested once they reach the bloodstream. The fate of each particular molecule from intake to final excretion from the body is reflected in EOs pharmacokinetics, which refers to several processes of bio-accessibility and bioavailability. EOs is mostly ingested by the skin, inhalation, and oral routes.

1.10. BIOAVAILABILITY OF EOS IN RELATION WITH ADMINISTRATION ROUTES AND EO ABSORPTION

EOs bioavailability is influenced by a variety of parameters, including physiochemical, biochemical, and physiological interactions. The monitoring of successive phases of EOs absorption, distribution, and excretion in the human body is required to understand their bioavailability.

The majority of investigations are conducted *in vitro* (e.g., Volic et al., 2018) or on animal models, and there is insufficient information on their behavior and fate in humans (e.g., Michiels et al., 2008; Zhang Y. et al., 2014). It is considered that intravenous administration of EOs has the highest bioavailability (100%) and that other administration routes have lower bioavailability.

However, as demonstrated for 1,8-cineole, which has a bioavailability rate of 95.6 percent, the bioavailability of EO components administered orally may be very high (Zimmermann et al., 1995).

Nonetheless, recent studies show that most EOs are swiftly absorbed whether administered by cutaneous, oral, or pulmonary routes. To estimate EOs bioavailability and, more importantly, bioactivity, it's crucial to understand how and in what amounts they reach the bloodstream, as well as how they're transported throughout the body, both of which represent EOs safety concerns (Tisserand and Young, 2014).

1.10.1. Dermal Administration

The outer and deeper dermis of the skin is separated by the stratum corneum, the outer epidermis layer, which serves as the first physical barrier to the penetration of exogenous substances (Godin and Touitou, 2007).

The intercellular (between skin cells), transcellular (through cells), and route through hair follicles, bypassing the stratum corneum, are thought to be the three probable skin penetration pathways (Williams and Barry, 2012).

The majority of EOs constituents are known to penetrate from the skin's surface, through the stratum corneum, into the dermis, and then into the bloodstream (Tisserand and Young, 2014). The high percutaneous absorption rates of EOs should be considered in risk evaluations in systemic toxicity because of its lipophilic characteristic.

Hydrophilic medications are better absorbed when combined with terpenes having polar functional groups, and lipophilic substances are better absorbed when combined with hydrocarbon terpenes (Godwin and Michniak, 1999). Lipophilic medications, on average, penetrate the skin barrier more thoroughly than hydrophilic drugs (Wester and Maibach, 2000) At concentrations of 5%, terpene compounds are also employed to promote transdermal medication administration (Aqil et al., 2007), owing to their great percutaneous enhancing ability and minimal cutaneous irritancy (Nokhodchi et al., 2007). EOs are also renowned for their beneficial effects on the skin, as well as the prevention and treatment of certain dermatological conditions (e.g., Sarkic and Stappen, 2018). EOs encapsulated in liposomes has recently been used in cosmetics (Sherry et al., 2013).

1.10.2. Respiratory Administration

Inhaled compounds travel down the trachea to the bronchi, then to the bronchioles, and lastly to the lung alveoli, which are excellent in transporting small molecules like terpenes into the bloodstream (Tisserand and Young, 2014). The beneficial effects of EOs on the respiratory system when inhaled are well-known and accepted at large (e.g., Maddocks-Jennings and Wilkinson, 2004).

1.10.3. Rectal and Vaginal Administration

When large systemic concentrations are required for bioactivity in the colon, rectal suppositories are employed. Because of the great sensitivity of the rectal mucosal membrane to EOs and the potential for irritation, dosage and concentrations should be carefully controlled (Tisserand and Young, 2014).

A similar conclusion applies to the vaginal application of EOs, which necessitates extra emulsification. Encapsulation of EOs with appropriate carriers for these purposes is likely to be developed in future trials.

1.10.4. Oral Administration

Dilution of free EOs in milk, soy milk, olive oil (Bilia et al., 2014), or other vegetable oils is commonly used for oral delivery. Oral ingestion, on the other hand, is the most popular method of using encapsulated EOs, notably in the case of food supplements and functional foods.

Entrapping EOs and/or their already extracted individual bioactive compounds (e.g., menthol, chamazulene, thymol, carvacrol, limonene, 1,8-cineole, etc.) enables targeted and controlled release, protection from degradation and losses, and masking of unpleasant taste and odor via oral intake (e.g., Bilia et al., 2014; Asbahani et al., 2015; Dajic Stevanovi ´ c et al., 2018 ´).

The stability of microencapsulated or nano-encapsulated carriers throughout the GI tract, as well as increased bioavailability and, specifically, the systemic activity of a specific medication at the action site, should all be considered when considering the bioavailability of the matrix and EO components.

The gastrointestinal tract is a tube-like structure with interconnected compartments that contain organs such as the liver, pancreas, and gallbladder. The oral cavity and salivary glands, the oesophagus, stomach, and small intestine (duodenum, jejunum, and ileum) make up the upper GI tract, while the large intestine (cecum, colon, and rectum) make up the lower GI tract (Treuting et al., 2018).

Four concentric layers surround the GI tract: the innermost mucous layer, which contains mucus and HCl secreting glands; a submucosal layer; and the outer muscular and outermost serous layers (Maderuelo et al., 2019).

The most serious side effect of delivering EOs or their constituents to any portion of the GI tract is irritation and inflammation of the mucosal membrane, which is dose-dependent (Tisserand and Young, 2014).

Nonetheless, many studies have highlighted the beneficial effects of EOs and their compounds on the gastric mucosa, with this gastroprotective activity linked to a variety of mechanisms, including a2-receptor activation, increased HSP-70, VIP, and PGE2 expression (heat shock protein, vasoactive intestinal peptide, and prostaglandin, respectively), and gastric SH group bioavailability (Rozza and Pellizzon, 2013).

According to Fernandes et al. (2012), lemongrass EO lowers stomach damage through endogenous prostaglandin pathways, whereas orange EO and limonene's gastroprotective activity is linked to an increase in gastric mucus formation (Moraes et al., 2009). As a result, in some situations, ensuring total EOs release from coating material in the stomach is feasible due to the favorable effects of bioactive substances on the gastric mucosa.

The stomach permits the absorption of water and various chemicals, including some specific terpenes, in addition to being responsible for digesting due to low gastric pH and pepsin's enzymatic activity (which should be less lipophilic, i.e., those containing polar groups, such as oxygenated monoterpenes, such as linalool, geraniol, neral, thymol, citronellol, etc.).

As per the investigations conducted *in vitro* and *in vivo* on piglet models, under *in vitro* settings, thymol, eugenol, carvacrol, and transcinnamaldehyde were weakly degraded in the proximal portions of the GI piglet tract. These EO components were virtually entirely absorbed in the upper GI tract, by the stomach, and the proximal small intestine, according to *in vivo* tests (Michiels et al., 2008).

Some EO elements can be broken down by digestive enzymes; for example, esters can be hydrolyzed in the stomach (Tisserand and Young, 2014). Because of the existence of villi and microvilli, the intestine as the primary absorption site has a large absorption area.

The rate of absorption of EO components is influenced by a number of parameters, including molecular weight, lipophilicity, solubility, and polarity (Esfanjani et al., 2018). As a result, various terpenes are absorbed at varying rates and by various regions of the GI tract.

In contrast to the quick absorption of thymol (Kohlert et al., 2002), the upper GI tract is not responsible for the absorption of 1,8-cineole (Kohlert et al., 2000), corroborating previous observations with ileostomy patients (Somerville et al., 1984). The quick absorption of terpenes was confirmed when 1,8-cineole and -pinene were detected in the plasma barely 30 minutes after oral administration (Zimmermann et al., 1995).

When EO components are mixed with pancreatic juice and bile salts in the duodenum, they are more easily soluble and absorbable (Michiels et al., 2008). The tiny size and lipophilic properties of EO molecules contribute to their rapid absorption (Kohlert et al., 2002). Most chemicals are absorbed in the GI tract through passive diffusion via trans-cellular and para-cellular absorption pathways, while some medications must be taken via carrier-mediated transport due to their membrane permeability reactions (Ho, 2011).

The stability of the capsule and, particularly, the bioactive substance, inside the stomach stage, is required for intestinal distribution and absorption. The bioavailability of medications is, in general, substantially influenced by GI pH variations.

The intraluminal pH varies along the GI tract, from very acidic in the stomach (lowest values in the digesting phase are 1–2 and about 4–5 during the resting phase) to neutral in the large intestine; pH values rise to 5.5 in the upper small intestine, reach a maximum of 7.5 in the ileocolonic region, then fall to 5.7 in the cecum, and rise again to pH of 6.7 in the rectum (Maurer et al., 2015).

It's reasonable to expect that GI absorption of various EO elements will follow the pH partition theory, with lipophilic, un-ionized chemicals passively transported across biological membranes (Maderuelo et al., 2019).

According to Abuhelwa et al., the stomach's acidic pH encourages the absorption of weak acids, whereas the small bowel's pH, which is closer to neutrality, facilitates the absorption of weak bases (2017).

Alcohols (and thus terpene alcohols) are thought to operate as both weak acids and weak bases, dependent on parameters including electronegativity, inductive and resonance effects, polarizability, and salvation as an external factor (Roberts and Caserio, 1977).

In this regard, terpene alcohols such as citral, linalool, geraniol, 1,8-cineole, and others may behave as acids or bases in the initial pH, depending on microenvironmental GIT circumstances, which will greatly influence their absorption location.

Terpenes having a carboxylic group (e.g., carnosic acid, abietic acid) and phenolic terpenes (e.g., thymol, anethol, carvacrol) on the other hand, act as weak acids and may be absorbed in the stomach, at least in part.

Finally, N-containing terpenes, such as amino acid derivatives other than L-phenylalanine and aminated terpenes, may operate as weak bases because they are more easily absorbed in the intestine.

Due to variable inter-molecular interactions, changes in molecule structure under varied GI tract circumstances, and unique properties of each individual body, absorption mechanisms are far more complex and cannot simply follow the pH partition theory. As a result, predicting the bioavailability of a specific EO element solely based on its physicochemical qualities, such as molecular mass, functional group, stability, solubility, pH responses, and so on, is problematic. These findings are insufficiently useful for hypotheses about EO interactions with a variety of biomolecules in complex biological systems such as human tissues and organs.

1.10.5. Metabolism, Distribution, and Excretion

To improve bioavailability, the bioactive chemical should have a high absorption rate and a low (renal) clearance rate, or excretion ability. When an EO component enters the bloodstream, the body begins to alter it, breaking it down into smaller, more polar molecules for easier filtration and removal via the kidneys (Djilani and Dicko, 2012).

Phase I (metabolism via the cytochrome P450 system) and phase II (glucuronidation, sulfation, and glutathione conjugation) of liver metabolism comprise oxidation and hydroxylation activities, as well as the addition of certain polar accessories. The chemical composition of EOs and their individual constituents have a significant impact on their metabolic destiny.

Phase II essential oil metabolites have been discovered in humans and animals in the form of glucoronides and sulfates, whereas expelled metabolites are primarily glycine and glucuronic acid, or are exhaled as CO2 (Kohlert et al., 2002). Terpenes are transported from the bloodstream to various tissues.

Their buildup is unlikely due to their fast clearance and short elimination half-life. It was discovered that 35 percent of the initial amount of menthol was renally excreted as menthol glucuronide, the predominant biliary metabolite capable of entering enterohepatic circulation, after oral administration (e.g., Kohlert et al., 2000; Grigoleit and Grigoleit, 2005). Thymol, carvacrol, limonene, and eugenol were shown to be comparable. Sulfate and glucuronide metabolites have been identified in urine and plasma after oral administration of EOs (Guénette et al., 2007; Michiels et al., 2008). Lipophilic chemicals, such as EO components, can cross the blood–brain barrier and interact with a variety of brain receptors, including glutamate and γ-aminobutyric acid receptors (Tisserand and Young, 2014). Monoterpenes and sesquiterpenes should only stay in the bloodstream for a brief time before being reallocated to muscle and then fat over time (Tisserand and Young, 2014). However, it is well known that EO compounds are rapidly absorbed, with just a small amount of the EOs remaining intact regardless of the method of delivery. The active components of EOs have a fast metabolism and a short half-life, therefore there is little chance of them accumulating in bodily tissues (Kohlert et al., 2002). The physicochemical qualities of EOs and their constituents, as well as their bioavailability and, in particular, biological effects at the target site, are all important factors in determining the suitable carrier system and encapsulation approach.

1.11. NEEDS FOR MICROENCAPSULATION OF EOS: ENCAPSULATION TECHNOLOGIES AND SELECTION OF CARRIER SYSTEMS

Microencapsulation of active compounds, such as EOs, serves two purposes:

1. it improves oxidative stability, thermostability, photostability, shelf life, and biological activity, as well as
2. ensures their transport to their intended location (Gallardo et al., 2013; Yang and McClements, 2013; Martins et al., 2014; Xiao et al., 2014; Yang et al., 2015)

Encapsulation of essential oils (EOs) regulates their volatility, sensory (primarily odor and taste), and release qualities, as well as their chemical stability and biological activity, under storage circumstances (e.g., Bilia et al., 2014). Encapsulation improves the bioactivity of EOs, according to several reviews (e.g., van Vuuren et al., 2010; Bilia et al., 2014; Dima et al., 2014; Asbahani et al., 2015; Li et al., 2015; Pandit et al., 2016; Maderuelo et al., 2019). When compared to free peppermint oil, encapsulated peppermint oil in starch-based emulsions had higher bioavailability, stability, and action against *Listeria monocytogenes* and *Staphylococcus aureus* (Liang et al., 2012). It was discovered that the antibacterial activity of EOs after nanoencapsulation typically outperformed that of existing antibiotics (Zaman et al., 2017).

The improved biological activity of encapsulated EO can be attributed mostly to its improved stability and reduced exposure to breakdown processes due to entrapment. However, it has been suggested that in some situations, such effects are the consequence of synergistic effects caused by EO interactions with carrier polymers as cashew gum or chitosan (e.g., Pandit et al., 2016).

(a) (b)

Figure 1.5. Peppermint Oil Sources.

Source: Bendifallah L, Tabli R, Khelladi H, Hamoudi-Belarbi L, Hamoudi S. Biological Activity of the Mentha spicata L. and Salvia officinalis L. (Lamiaceae) Essential Oils on Sytophilusgranarius L. and Triboliumconfusum Jac. Du Val. Infested Stored Wheat. Biology and Life Sciences Forum. 2021; 4(1):108. https://doi.org/10.3390/IECPS2020-08873

Finally, nanometric particle encapsulation contributes to improved cellular absorption mechanisms and bio efficacy (Bilia et al., 2014). Encapsulation technology has a significant impact on the stability, release kinetics, and related bioavailability and bioactivity of encapsulated materials.

Depending on the fundamental properties of the core material and the carrier, as well as the use of the encapsulated material, a variety of encapsulation processes, such as spray drying, extrusion, coacervation, emulsification, and so on, are utilized nowadays for encapsulation of EOs.

Spray drying is a fast, continuous, and relatively low-cost industrial procedure with simple scale-up. Maltodextrins, starch, gum Arabic, and chitosan are some of the biomolecules used as spray-drying carriers (Ersus and Yurdagel, 2007; Kausadikar et al., 2015). The following are some of the disadvantages of the spray-drying procedure in general:

1. Non-uniform particles' size and shape,
2. Tendency of particles to aggregate,
3. Carrier material solubility in water at an acceptable level, and
4. Short-time exposure to high temperatures.

For covering EO droplets by a single shell (simple coacervation) or two-layer shell (complex coacervation), coacervation could be suitable technique. Shell materials should ensure:

1. The rigidity of oil carrier,
2. Thermal stability, and

3. Chemical stability under gastric conditions and
4. Should be dissolved in the intestinal fluid.

The inner layer, which is closest to the oil droplets, and the outer layer make up the two-layer shell. Amphiphilic materials, such as proteins or certain surfactants, can be used to create the inner layer (Torcello-Gomez et al., 2011).

Polysaccharide hydrogels are commonly used on the outer layer to maintain the mechanical stability of oil carriers and to meet process parameters. Extrusion is a technique that can be combined with coacervation to create a new product. This method is appropriate for preparing

1. Polysaccharide monophase and multiphase hydrogel matrices, and
2. Polysaccharide–protein hydrogel blends in the form of microbeads (Nedovic et al., 2011; Voli ′ c et al., 2018; Obradovic ′ et al., 2019).

The advantages of extrusion are

1. The chemical stability of beads under storage conditions and gastric conditions,
2. The mechanical stability of the beads, and
3. The possibility of encapsulation of hydrophobic or hydrophilic active compounds.

The disadvantages of this technique could be

1. low production rate and
2. Scale-up difficulties (Gouin, 2004).

Emulsification is another commonly used method. Emulsions, such as vegetable oils with added proteins and emulsifiers, are a viable choice for encapsulating hydrophobic active chemicals as carriers.

In comparison to extrusion, the emulsification process is better for creating small-sized particles (10 m1 mm). Processing costs, on the other hand, appear to be higher than for extrusion. The choice of carrier material for oral administration depends on

1. The surface activity of the active compound
2. Processing conditions
3. Storage conditions, and
4. Cost and scale of production.

When it comes to delivering carriers to the lower intestine, they must maintain their integrity under stomach conditions during the retardation period. It is vital to examine the following factors in order to enhance carrier performance.

1. The interactions of pepsin with the main constituents of the carrier's matrix at the molecular level and
2. Carrier structure at the supramolecular level.

Carriers, often used in the literature, could be classified into two groups:

1. Carriers made from natural macromolecules and
2. Lipid-based carriers.

Carriers made by natural macromolecules can be divided into a few groups:

1. Monophase polysaccharide hydrogels
2. Multiphase polysaccharide hydrogels in the form of blends or multilayer microbeads, and
3. Polysaccharide– protein hydrogels in the form of blends
4. Lipid-based carriers, such as some vegetable oils and liposomes, and
5. Lipid–protein carriers.

Various polysaccharide ionic hydrogels have been used as the matrix, including Ca–alginate (Chan, 2011), alginate–cashew gum (de Oliveira et al., 2014), cashew gum–inulin (de Barros Fernandes et al., 2016), alginate–xanthan gum (Zhang S. et al., 2014), xanthan gum–pectin (Qiu al., 2015), alginate–pectin (Wang et al., 2013), and alginate–chitosan (Xu et al., 2007), have been utilized as the carrier's matrix primarily for the entrapment of hydrophilic active compounds, such as some types of polyphenols. Because of the amphiphilic characteristics of proteins, polysaccharide–protein hydrogels are appropriate for the trapping of hydrophobic active chemicals such as EOs and some types of polyphenols. Lupin (Piornos et al., 2017), soybean lecithin (Torcello-Gomez et al., 2011), gliadin (Qiu et al., 2015), whey proteins (Zhang et al., 2016), gelatin (Roy et al., 2009), and many other proteins are commonly combined with alginate, pectin, and xanthan. Liposomes can be used to encapsulate both hydrophilic and hydrophobic active ingredients (Akbarzadeh et al., 2013). Because of their resistance to pepsin action, vegetable oils derived from long-chain triglycerides are a viable choice for encapsulating EO (Yara-Varon et al., 2017).

1.11.1. Polysaccharide-Based Carriers

Polysaccharides including alginate, chitosan, and maltodextrin are commonly utilized to encapsulate EOs in the form of physical or chemical hydrogels (Ravichandran et al., 2014; GomezMascaraque et al., 2015; Pasukamonset et al., 2016).

Alginate is a generic term for a group of water-soluble, linear macromolecules with a large molecular mass (between 32,000 and 400,000 g/mol) derived primarily from brown algae species.

The distribution of mannuronic acid and guluronic acid units in alginate chains of G-G, M-M, and M-G residue blocks [where G is α-L-gluronic acid, and M is β-D-mannuronic acid, (Milivojevic et al., 2015)], as well as their ratio (M/G), are mostly determined by the natural source of alginate (algae type, season, location, etc.) and dictate the physical and chemical properties.

Chitosan is primarily made up of (1, 4) connected 2-amino-2-deoxy-D-glucan chains. Chitosan chains act similarly to alginate chains in that they are semiflexible. At pH 7.0, anion-like alginate chains form a gel with cation-like chitosan chains, with varying rheological behavior depending on the alginate-to-chitosan ratio. Partial hydrolysis is used to remove maltodextrin from starch.

D-glucose units are linked in chains of varying lengths in this form of polysaccharide. Under simulated GI circumstances, Ca–alginate has been proposed to improve bioavailability, thermal stability, and biological activity of active substances.

However, weak interactions between active chemicals and the hydrogel matrix may result in undesired EO leakage. Ca–alginate beads could be coated with chitosan to boost the carriers' performance. The development of alginate–chitosan complexes provides extra carrier stability (Popa et al., 2000; Anbinder et al., 2011).

Devi and Maji (2009) encapsulated neem EO in chitosan–carrageenan carriers. Neem seed oil, commonly known as margosa oil, is a marketed product obtained from the fruits of the neem tree. Antimicrobial action against *Candida utilis*, *Bacillus cereus*, and *Bacillus subtilis* is ensured by microencapsulating pimento EO in chitosan and k-carrageenan (Dima et al., 2014).

Gum Arabic was utilized by Dong et al. (2011) to microencapsulate peppermint essential oil. Encapsulation ensures that peppermint EO (and its primary constituents, primarily menthol and isomenthol) is released for a long time (Sarkar et al., 2013).

Maltodextrin is a hydrolyzed starch that is commonly used for microencapsulation of essential oils (EOs) in combination with surface active biopolymers such as gum Arabic (Fernandes et al., 2008; Bule et al., 2010; Kausadikar et al., 2015), modified starches (Bule et al., 2010), and proteins (Hogan et al., 2003;

Bae and Lee, 2008). It is due to the fact that maltodextrin has high thermal stability and provides oxidation resistance.

This polysaccharide, however, has a low emulsifying capacity. As a result, it is preferable to combine maltodextrin with surface-active biopolymers in order to improve bioactive compound volatile retention during the drying process (Ersus and Yurdagel, 2007; Fang and Bhandari, 2010; Paz et al., 2010; Mahadivi et al., 2016; Tolun et al., 2016).

To boost the stability of components like 1,8-cineole and α-terpinyl acetate, a composite consisting of gum Arabic, modified starch, and maltodextrin was employed for microencapsulation of cardamom EO (Krishnan et al., 2005). For microencapsulation of cumin oleoresin, Kanakdande et al. (2007) utilized comparable carriers.

Cumin volatile EO is made up of terpenes such as β-pinene, p-cymene, and γ-terpinene), aldehydes (cuminaldehyde, 1,3-pmentha, and 3-p-menthen-7-al), and terpene alcohol, according to the researchers. The dextrose equivalent determines the oxygen barrier characteristics of maltodextrin (DE).

Better DE carriers ensure more intense interactions between active chemicals and matrix, resulting in higher encapsulation efficiencies. These carriers are less oxygen permeable. (Touré et al., 2011; Simon-Brown et al., 2016) have effectively used high DE maltodextrin in the encapsulation of lemon EO, orange peel EO, cardamom EO, and ginger EO to protect them from oxidation.

Recently, it was discovered that maltodextrin, when used as a carrier material, can shield polyphenol chemicals from enzyme action in simulated GI settings (Romano et al., 2017). Antioxidants and phenolic chemicals have been transported via xanthan gum, an extracellular microbial polysaccharide (Da Rosa et al., 2013; Rutz et al., 2013).

Xanthan gum, when combined with maltodextrin and chitosan as a carrier material, has the capacity to establish strong electrostatic contacts between the amino groups of chitosan (polycation) and the carboxylic groups of xanthan gum (polyanion).

Researchers have shown that this sort of mix carrier improves the regulated release rate of the encapsulated substance (Martínez-Ruvalcaba et al., 2007; Da Rosa et al., 2013). The biopolymer complexes of xanthan gum and whey proteins were also found to have the emulsifying capacity and a good influence on the controlled release of water-soluble nutraceuticals in W/O/W double emulsion delivery systems.

1.11.2. Protein-Based Carriers

Because of their binding hydrophobic contacts and hydrogen bonding attraction between molecules, proteins as natural food-grade polymers were employed as carrier materials for microencapsulation of various EOs: (1) alone and (2) in combination with polysaccharide hydrogel (Zou et al., 2012; Haratifar and Corredig, 2014; Chuacharoen and Sabliov, 2016).

Self-cross-linking of protein interchain and intrachain is required for the formation of a protein carrier matrix. In some circumstances, crosslinking can be produced by heat treatment or changing the pH of a solution (Shpigelman et al., 2010; Tavares et al., 2014).

Encapsulation of thermosensitive, hydrophobic bioactive substances has traditionally been done with protein-based carriers such as gelatin, casein, whey proteins, and soy proteins (Pool et al., 2013; Xue et al., 2014; Jia et al., 2016).

The binding affinity of polyphenols and EOs to the protein matrix determines the encapsulation effectiveness of these carriers (Livney, 2010). The ability of bioactive chemicals to be released from protein-based transporters is affected by pH. At neutral pH, significant swelling of protein-based transporters could result in unwanted bioactive chemical leakage (Kimpel and Schmitt, 2015; Liu et al., 2015).

Key breakdown under stomach conditions due to pepsin attack is a significant disadvantage of this type of carrier (Kumar et al., 2016). Gelatin was employed by Sutaphanit and Chitprasert (2014) to microencapsulate basil essential oil. Methyl eugenol (42.58 percent), caryophyllene (26.88 percent), and eugenol are the main constituents (10.66 percent).

When compared to protein-based carriers, a combination of protein–polysaccharide carriers can improve delivery system mechanical and release features while also preventing enzymatic degradation of proteins in stomach conditions (Diaz-Bandera et al., 2013; Jia et al., 2016). Furthermore, globular proteins, as well as whey protein hydrolysates, have been shown to have antioxidant activity in numerous studies. They can minimize the unfavorable oxidation of EOs and improve the carriers' oxidative stability (Dryakova et al., 2010; Carneiro et al., 2013).

Encapsulating EOs in protein–polysaccharide carriers is a good way to (1) stabilize these active chemicals and (2) protect them from chemical destruction (Turasan et al., 2015; Campelo et al., 2017).

Tween 80 was used to stabilize the oregano EO emulsion, which was then encapsulated in a variety of microcarriers, including milk powder and whey protein particles, rice starch particles and inulin, and gelatin–sucrose composite (Beirao da Costa et al., 2012).

The nutritional quality, ease of preparation, and low cost of polysaccharide carriers are all advantages. Low encapsulation efficiency, loading capacity, and release efficiency in the small intestine are all disadvantages (de Oliveira et al., 2014).

The addition of protein clusters increases the thermal and mechanical stability of the protein as well as the nutritional value. Low encapsulation efficiency, loading capacity, and release efficiency in the small intestine are other disadvantages of polysaccharide/protein carriers for encapsulation of EOs (Dajic Stevanovi'c et al., 2018; Voli'c et al., 2018).

1.11.3. Lipid-Based Carriers

Widely used for encapsulation of EOs, Lipid-based carriers made by vegetable oils have the following advantages in the capacity of a carrier (Bilia et al., 2014)

1. Good encapsulation efficiency,
2. Stability of active compounds under storage conditions and under gastric conditions, and
3. Thermal stability (Campos et al., 2014).

The mixture of Triglycerides (major components) and minor components (5%), such as glycerolipids, such as monoglycerides and diglycerides, phospholipids, and non-glycerolipids, such as sterols, tocopherols/ tocotrienols, free fatty acids, vitamins, pigments, proteins, phenolic compounds, water, and so on, make up vegetable oils (YaraVaron et al., 2017).

Liposomes and solid lipid carriers are two more forms of lipid-based particles. Hydrophobic, hydrophilic, and amphiphilic compounds can all be encapsulated using this sort of carrier (Yoshida et al., 2010).

For the manufacture and encapsulation of EOs, thin film hydration, freeze–thaw, sonication, and reverse-phase evaporation were the most often utilized procedures. Liposomes have the following drawbacks:

1. Complex and expensive preparation procedures and
2. Reduced stability under storage conditions that could restrict their applications (Akbarzadeh et al., 2013).

For the delivery of polyphenol-type catechins (EGCG), solid lipid nanoparticles and nanostructured lipid carriers were utilized as nanocarriers (Shi et al., 2013). Proteins and lipid carriers can be coupled to improve chemical and mechanical stability, as well as encapsulation efficiency.

Polyphenols have been successfully encapsulated using carriers created by combining proteins and lipid components (β-lactoglobulin-medium chain triglyceride) (Pool et al., 2013). When compared to polysaccharide/protein carriers, lipid carriers have superior encapsulation efficiency, loading capacity, and release efficiency in the small intestine.

However, limited mechanical and thermal stabilities, as well as a difficult manufacturing technique and higher cost when compared to polysaccharide/protein carriers, are disadvantages.

1.12. CONCLUSION

The chapter discussed the basic concept of macromolecules. It also discussed synthetic polymers and biological polymers. This chapter discussed the science behind the macromolecules and the distribution of the molecular weight. It discussed macromolecular thermodynamics. Towards the end of the chapter, it explained the natural macromolecules as carriers for essential oils – from extraction to biomedical applications.

REFERENCES

1. Chen, L., Yang, Z., McClements, D., Jin, Z. and Miao, M., 2022. Biological macromolecules for nutrients delivery. *Biological Macromolecules*, [online] pp.455-477. Available at: <https://www.sciencedirect.com/science/article/pii/B9780323857598000208?via%3Dihub> [Accessed 14 June 2022].
2. Dajic Stevanovic, Z., Sieniawska, E., Glowniak, K., Obradovic, N. and Pajic-Lijakovic, I., 2020. Natural Macromolecules as Carriers for Essential Oils: From Extraction to Biomedical Application. *Frontiers in Bioengineering and Biotechnology*, [online] 8. Available at: <https://www.ncbi.nlm.nih.gov/pmc/articles/PMC7330110/pdf/fbioe-08-00563.pdf> [Accessed 14 June 2022].
3. Feng, X., 2022. PLANT - DERIVED MACROMOLECULES IN THE SOIL. *Multi - Scale Biogeochemical Processes in Soil Ecosystems*, [online] pp.79-115. Available at: <https://onlinelibrary.wiley.com/doi/10.1002/9781119480419.ch4> [Accessed 14 June 2022].
4. Liu, Q. and Urban, M., 2022. Stimulus-Responsive Macromolecules in Polymeric Coatings. *Polymer Reviews*, [online] pp.1-35. Available at: <https://www.tandfonline.com/doi/abs/10.1080/15583724.2022.2065299?journalCode=lmsc20> [Accessed 14 June 2022].
5. SUN, S., 1992. *PHYSICAL CHEMISTRY OF MACROMOLECULES*. 2nd ed. [ebook] JOHN WILEY & SONS, INC. Available at: <https://unpa.edu.mx/~aramirez/Physical%20Chemistry%20of%20Macromolecules%20Basic%20Principles%20and%20Issues.pdf> [Accessed 14 June 2022].

CHAPTER 2

Cisplatin Derivatives as Antiviral Agents

CONTENTS

The attention is once again on the critical need for antiviral drug research to address both old and developing virus epidemics. The recent attempt to restart the smallpox vaccination program to combat biological terrorism has been halted, in part due to the observation of adverse cardiac events following vaccination and the Centers for Disease.

2.1. INTRODUCTION

Centers for Disease Control has recommended that people at risk for heart disease be excluded from the vaccination program. Even though the quantity and intensity of adverse responses to most, if not all, vaccinations do not appear to be growing, the fraction of the human population vulnerable to these issues does appear to be increasing.

Increased lifespan, as well as viral and environmental variables, are creating a population of immunocompromised people for whom immunizations are ineffective. Diseases that are caused by ancient adversaries, like influenza, have always had the potential to produce new pandemics.

Unlike vaccinations, antivirals have the potential to function against new variations of existing viruses as well as novel viruses such as SARS and avian flu, as well as interventions in biological terrorism. Furthermore, a new attitude seems to be evolving wherein vaccination dangers are widely accepted, with the rotavirus vaccine being the most recent victim.

Polymeric medications may have several benefits over small molecule pharmaceuticals now in use. Polymeric pharmaceuticals can either operate as a drug or as a way of releasing the substance over time. Controlled or extended drug release allows for a lower concentration of the medication to be present at any one moment in ordinarily sensitive locations like the kidneys and liver while increasing the quantity of drug staying within the target site (s).

Tumor-associated vasculatures, for example, are typically permeable to plasma proteins and other macromolecules. Both leaky vasculatures and restricted lymphatic drainage, which are characteristic of tumors and absent in normal tissue, lead to an accumulation of macromolecules such as polymeric medicines in the interstitial space of a wide range of malignancies.

This one is referred to as the "enhanced permeability and retention effect," or EPR. Trapped polymeric medicines can thus act as polymeric drugs for a longer period of time or in controlled release of the active substance. This continual release of bioactive medications is essential for cell-cycle-specific therapeutics and also drug resistance reduction.

Nevertheless, all antivirals suffer from the effects of target specificity. Viruses, for the most part, replicate the viral genome and make new virus particles using cellular machinery. In an attempt to attack viral replication, uninfected cells' biological activities are also adversely impacted. A few of these side effects can be avoided using polymeric medicines.

There has been an investigation of a variety of metal-containing polymers as possible antiviral drugs, with a focus on platinum and organotin-containing polymers. Polymeric medications have several potential benefits over monomeric or small molecule therapies.

Researchers have suggested that at least some cancers have a viral relationship. Thus, we have begun testing polymers that show good anticancer activity against a variety of viruses.

Experts suspect that at least some malignancies are linked to viruses. As a result, scientists recently started to implement testing polymers with anticancer activity against a number of viruses. Researchers recently examined a variety of organotin products made from well-known antibacterial medicines, including ciprofloxacin, ampicillin (Figure 2.1.), and norfloxacin (Figure 2.2.). All of these drugs showed antiviral activity against a variety of viruses, including reovirus ST3, vaccinia virus, herpes simplex virus (HSV-1), and varicella zoster virus (VZV). At doses of around 2 mg/ml, the organotin polymers of norfloxacin and ampicillin inhibited virus growth completely, but neither norfloxacin nor ampicillin inhibited virus growth.

Organotin derivatives of the well-known antiviral medication acyclovir were also assessed. Again, the organotin polymers inhibited the viruses at lower concentrations than acyclovir alone.

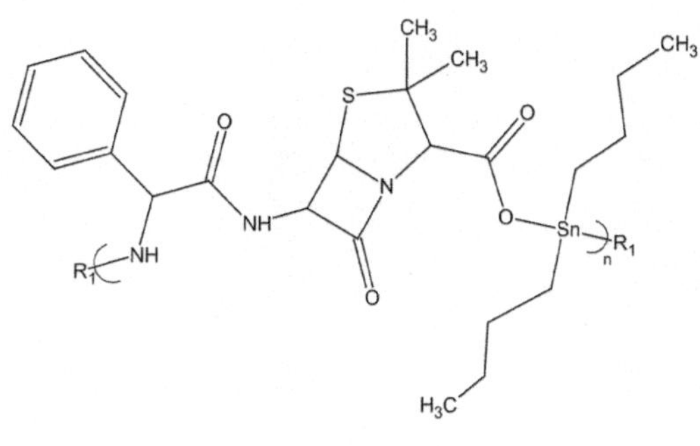

Figure 2.1. Dibutyltin-ampicillin polymer.

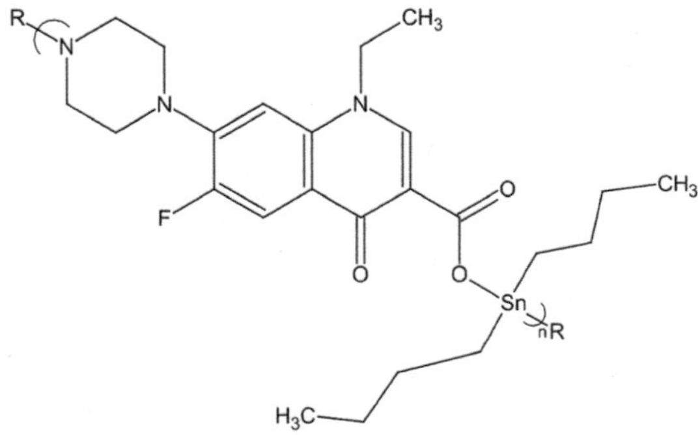

Figure 2.2. Dibutyltin-norfloxacin polymer.

As a result, the combination of recognized medicines with organotin moieties inside polymers appears to be more potent antiviral agents than either of the reactants alone.

The interest in these organotin-containing polymers stems from the fact that they are strong anticancer drugs capable of suppressing cancer cell proliferation at doses comparable to or lower than cisplatin. Furthermore, these organotin polymers are far less hazardous than the most commonly used anticancer medication, cisplatin.

For over 30 years, researchers have been researching a variety of polymeric cisplatin derivatives as anticancer medicines. Once more, scientists have produced polymeric medicines that limit cancer cell growth at concentrations comparable to cisplatin, and these compounds are significantly less hazardous. Some of these polymeric cisplatin derivatives have also been studied as antiviral agents.

2.2. INHIBITION

The vast majority of viruses are RNA viruses. It is not unexpected that they are also responsible for the vast majority of human ailments. Some of the most well-known illnesses caused by RNA viruses are as follows:

- Common Cold
- Poliomyelitis
- Hepatitis
- Encephalitis
- Yellow Fever
- Rubella

- Influenza
- Measles
- Mumps
- Various Hemorrhagic Fevers

2.2.1. Features of an Ideal Antiviral Drug

An ideal antiviral medicine would have the following characteristics: efficient suppression of a critical viral function, a mechanism to prevent the development of drug-resistant viruses, broad-spectrum effectiveness against RNA and DNA viruses, and no deleterious influence on host cell activities.

2.2.2. Strategies for Antiviral Therapy

Many antivirals aim for one of five main viral procedures:

1. virus attachment to the host cell,
2. virus penetration and/or uncoating to start releasing the viral nucleic acid into the host cell,
3. viral genome replication,
4. viral gene expression to start producing viral proteins, and
5. virus structure assembly and maturation and release of progeny virions with or without host cell lysis.

2.2.3. Attachment

Virus attachment can be prevented in two ways, which are discussed further below.

1) Viral attachment protein (VAP) mimicking agents can be introduced into the infected host. These VAPs subsequently connect to the cellular receptor and prevent the virus from binding. Anti-idiotypic antibodies that imitate the VAP can be generated. When these antibodies are delivered to the host, they attach to the cellular receptors that infectious viruses would typically have access to.

This "blocks" the viral receptor on the cell, preventing the virus from adhering and infecting it. Natural ligands of the viral receptor can be used to bind the receptor and prevent the virus from using it. Vaccinia virus and the epidermal growth factor (EGF) receptor are two examples of this.

The fourth option is to utilize synthetic ligands that are similar to the VAP's receptor-binding domain. These peptides would attach to a cell receptor and inhibit the receptor from binding to the VAP, so preventing infectious disease.

2) Agents that operate by binding the VAP and imitating the viral receptor on the host cell. Antibodies against the VAP, which are spontaneously created in response to most viruses, bind the VAP and block it from interacting with host cell receptors. Anti-idiotypic antibodies, which imitate the cell receptor or external receptors utilized by HIV, can be generated.

When these antibodies are injected into the host, they act as "binding targets" for the virus, but unlike the cells that typically produce these receptors, they cannot be infected or promote viral reproduction. Furthermore, synthetic receptor mimics can be created to bind virus before it has a chance to contact cell receptors. The use of sialic acid derivatives to bind influenza virus is one example of this strategy.

2.2.4. Penetration and Uncoating

Several viruses' molecular mechanisms have been challenging to investigate, making it difficult to particularly target certain phases of the viral life cycle. Uncoating is primarily mediated by cellular enzymes, but it is frequently modified by one or more viral proteins, as is penetration.

Pleconaril is an anti-picorna virus agent with a broad spectrum of activity. It is a tiny cyclic medication that binds to the virus's capsid pore. As a result, it prevents the viral particle from attaching and uncoating.

Amantadine (Figure 2.3.) and rimantadine (Figure 2.4.) are both antiviral drugs that work against influenza A viruses. The mechanism of action of these closely related drugs is complicated and unknown, however, they are thought to disrupt cellular membrane ion channels. Both medications work by targeting the influenza a matrix protein (M2). Drug-treated cells are often unable to reduce the pH of the endosomal compartment (a function typically regulated by the M2 gene product), a step required to cause conformational changes in the HA protein to allow membrane fusion.

2.2.5. Genome Replication

Most viruses have developed their own enzymatic methods to redirect cellular energy to viral genome replication. There are frequently enough variations between viral and cellular polymerases to give a target for an antiviral medication while causing no harm to the unaffected agents. This method produced the vast majority of antiviral medicines now in use.

The majority of these medications work as polymerase substrates, or as nucleoside/nucleotide analogs.

The toxicity of these medications varies greatly, from those that are well tolerated, such as acyclovir, to others that are extremely toxic, such as IdU/TFT and AZT. The pharmacokinetics of these nucleoside analogues is problematic, as they often have short serum half-lives of 1 to 4 hours.

Nucleoside analogues are pro-drugs, which means they must be phosphorylated before they can be activated. Acyclovir (Figure 2.5.) is phosphorylated 200 times faster by the herpes simplex virus (HSV) thymidine kinase than by cellular enzymes. Gancyclovir (Figure 2.5.) is 10 times more potent than acyclovir against cytomegalovirus (CMV) because it is selectively phosphorylated by a CMV-encoded enzyme not seen in HSV.

Figures 2.6. – 2.14. show further nucleoside analogs generated from Acyclovir and Gancyclovir that are active against herpesviruses. Figures 2.15. – 2.17. demonstrate more nucleoside analogues with anti-HIV activity.

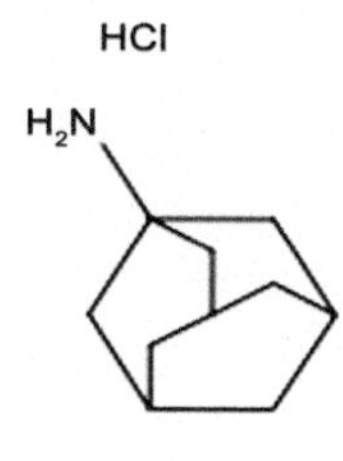

Figure 2.3. Amantadine.

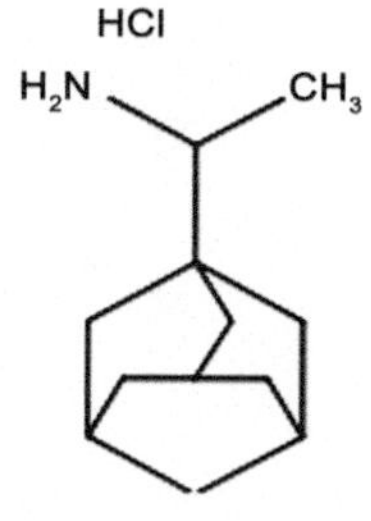

Figure 2.4. Rimantadine.

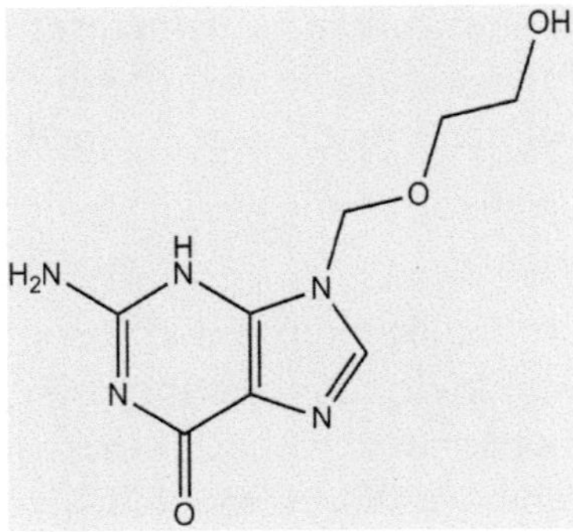

Figure 2.5. Acyclovir.

2.2.6. Gene Expression

Several viruses rely extensively on cellular machinery for viral genome transcription, mRNA splicing, translation, and protein trafficking. Unlike genome replication, distinctively viral proteins are not engaged in these activities, and none have been used as antiviral therapeutic targets yet.

Assembly, Maturation, and Release of Progeny Virus

The mechanisms of viral assembly, maturation, and release of progeny virus are poorly known for the vast majority of viruses. There are two anti-influenza medications on the market. These are Relenza in the form of an aerosol and Tamiflu in the form of a tablet. Tamiflu has antiviral activity both against influenza A and B viruses. Both of these medications act as neuraminidase inhibitors, preventing the release of budded viruses from cells.

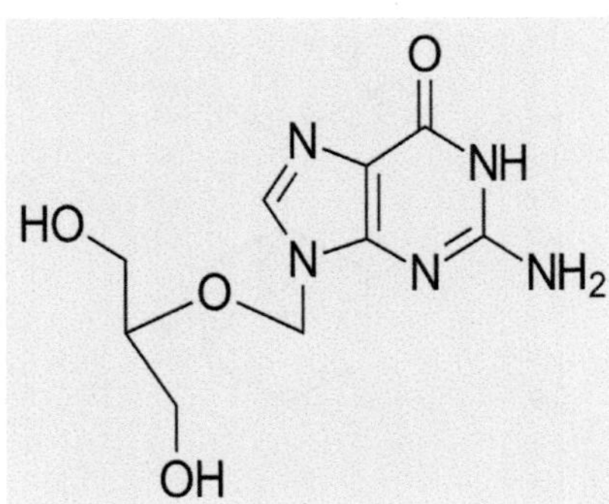

Figure 2.6. Gancyclovir.

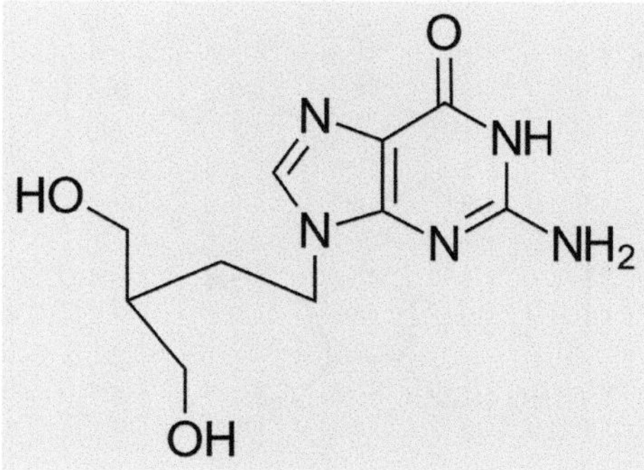

Figure 2.7. Penciclovir.

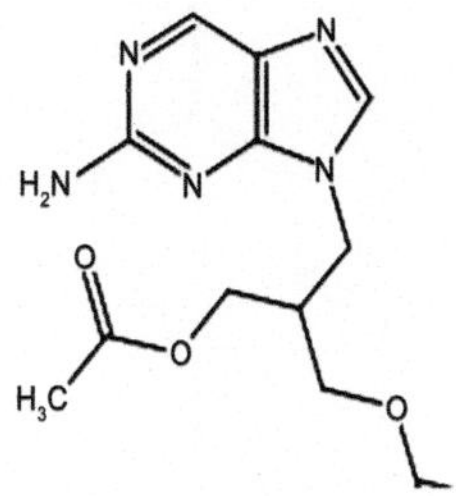

Figure 2.8. Famciclovir.

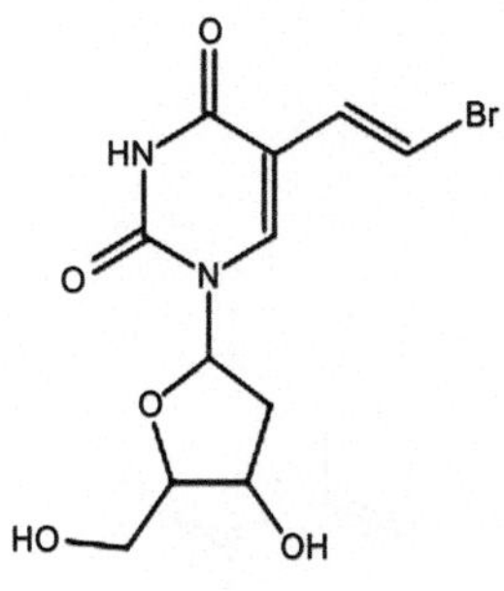

Figure 2.9. BVDU.

2.2.7. Additional Antiviral Drugs

If ganciclovir medication is unsuccessful or poorly tolerated, foscarnet (Figure 2.18.) is a first-line treatment for CMV retinitis and treatment for CMV colitis. Foscarnet may cure sensitive infections in the brain since it passes the blood-brain barrier.

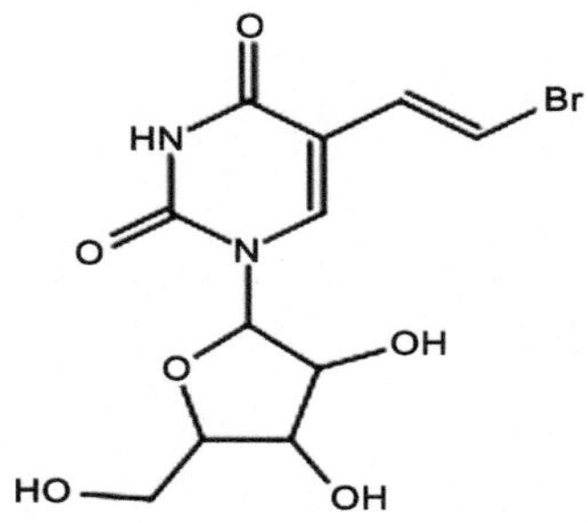

Figure 2.10. Brovavir,

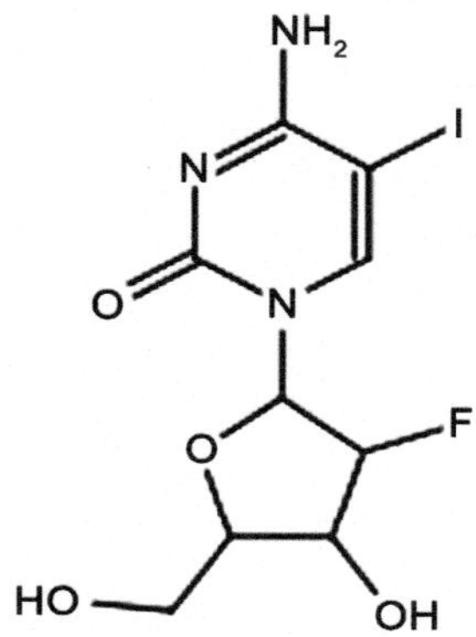

Figure 2.11. FIAC.

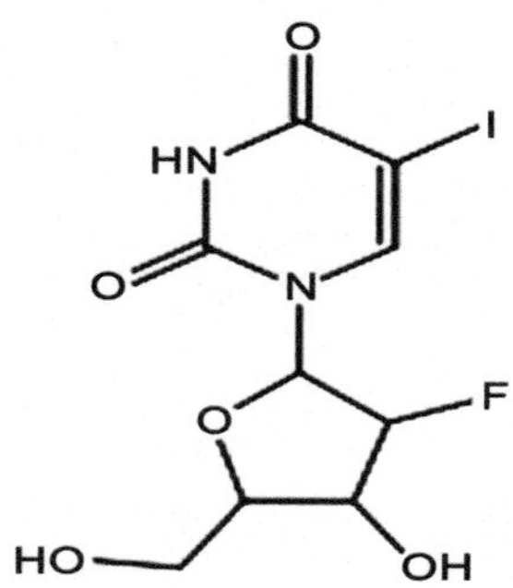

Figure 2.12. FIAU.

Foscarnet could be used to treat herpes strains that are resistant to acyclovir therapy. Idoxuridine (IdU) (Figure 2.19.) functions by permanently replacing thymidine in newly generated DNA, resulting in an aberrant, basically nonfunctional DNA molecule. The medication is very hazardous to host cells and acts on viral and host cell DNA.

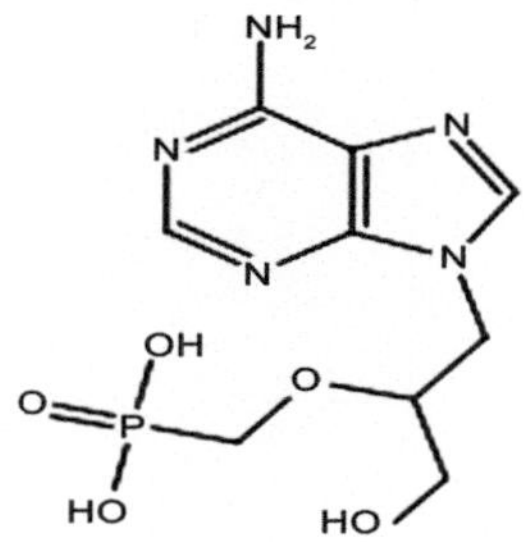

Figure 2.13. (S)-HPMPA.

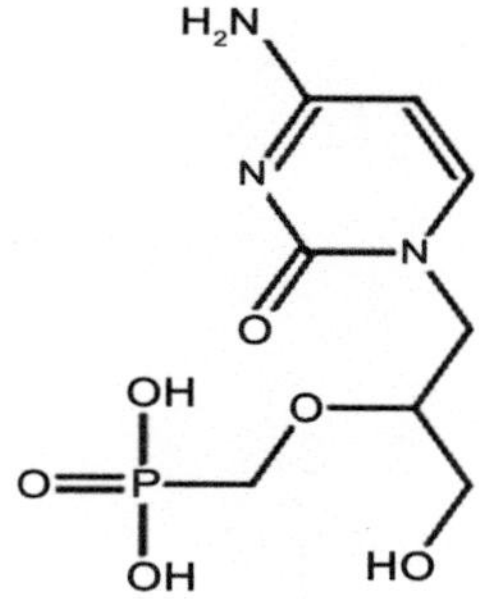

Figure 2.14. (S)-HPMPC.

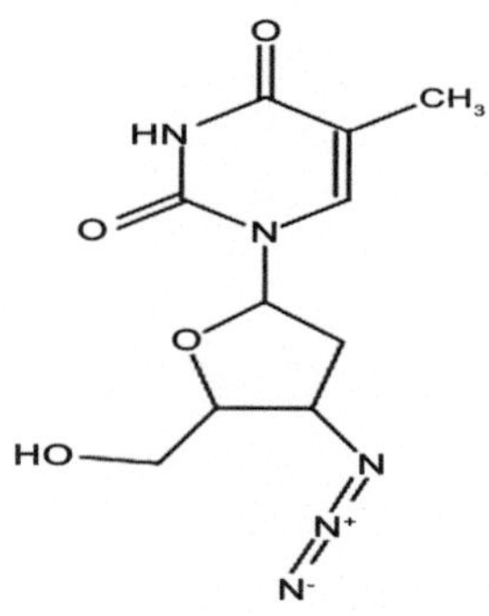

Figure 2.15. AZT.

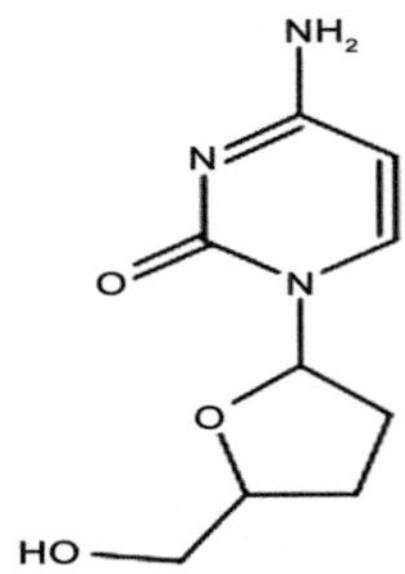

Figure 2.16. ddC.

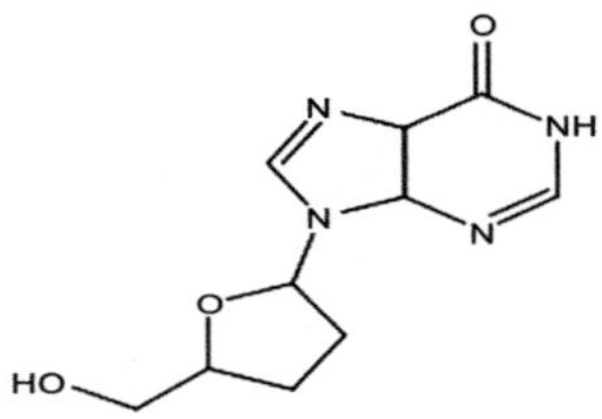

Figure 2.17. ddI.

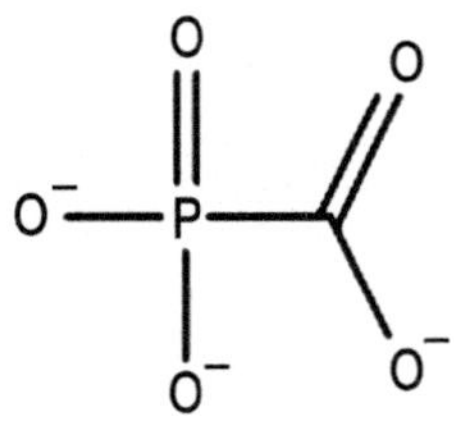

Figure 2.18. Foscarnet.

IdU has been restricted to topical treatment of herpes simplex keratoconjunctivitis due to its severe systemic toxicity. Ribavirin is a guanosine analogue that prevents many RNA and DNA viruses from replicating.

Ribavirin (Figure 2.20) is considered to prevent the synthesis of messenger RNA. Ribavirin inhibits respiratory syncytial virus (RSV), influenza A and B, HSV-1, HSV-2, and many other viruses *in vitro*. Vidarabine (adenine arabinoside, ara-A) (Figure 2.21) inhibits viral DNA synthesis and is used to treat HSV infections.

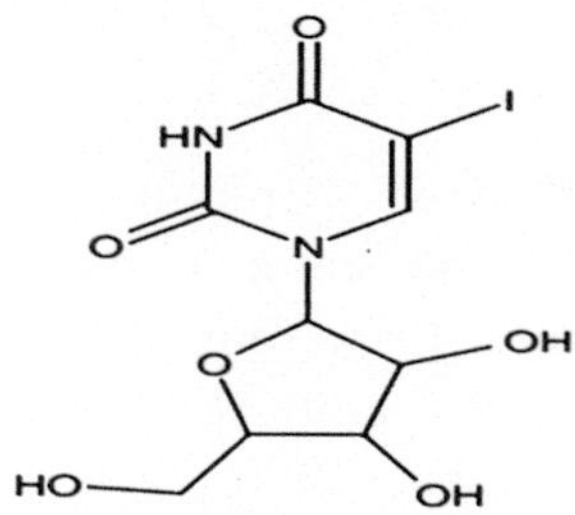

Figure 2.19. IdU.

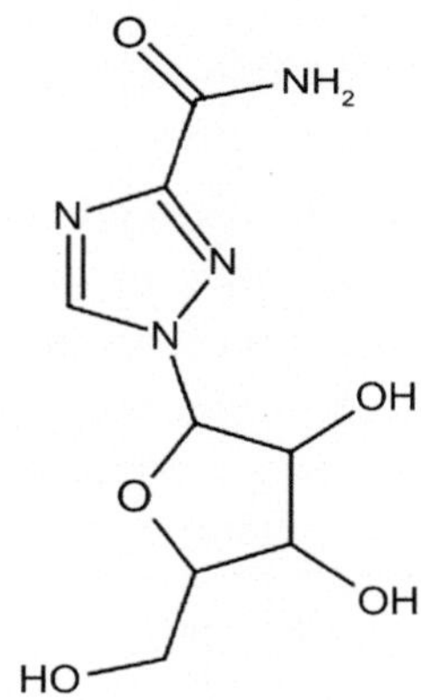

Figure 2.20. Ribavirin.

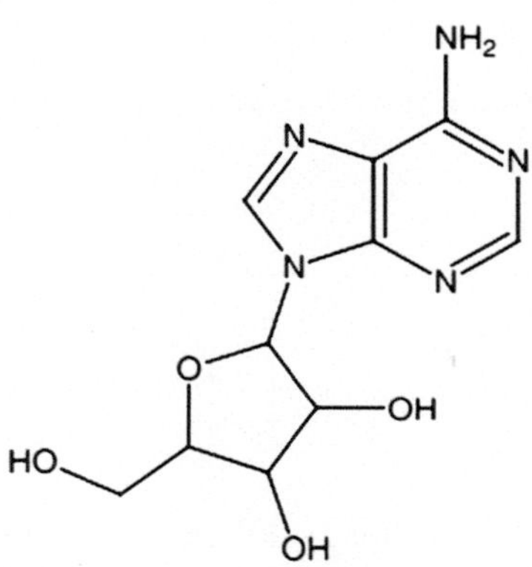

Figure 2.21. Vidarabine.

2.3. CURRENTLY APPROVED PLATINUM-CONTAINING DRUGS

Despite the fact that hundreds of cisplatin analogs have been made and tested, only roughly 28 platinum compounds have been tested in clinical trials as anticancer medicines. Only four of these are presently approved. Cisplatin (Figure 2.22.), carboplatin (Figure 2.23.), oxaliplatin (Figure 2.24.), and nedaplatin have all been authorized (Figure 2.25.). Only the first two are commercially accessible for use in cancer therapy.

Carboplatin (Figure 2.23.) is the most often used metal-containing anticancer medication after cisplatin (Figure 2.22.). Although its cell-killing power is similar to that of cisplatin, it has shown modest efficacy in some malignancies that are less susceptible to cisplatin, such as non-small cell lung cancer. It also has a distinct pharmacokinetic profile.

The presence of the bidentate carboxylate moiety results in slower reaction rates with the biological environment. As a result, it has reduced nephrotoxicity and is favored for people with renal failure. It also demonstrates a lower rate of serum protein binding, with just 10 to 20% permanently bound to protein. This increases the bioavailability and concentrations of the delivered medication by around a factor of five.

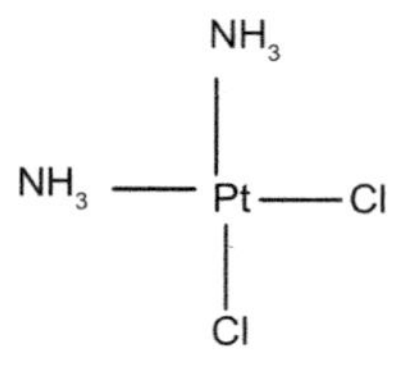

Figure 2.22. Cisplatin, Cis-DDP (Cis-diaminedichloroplatinum(II)).

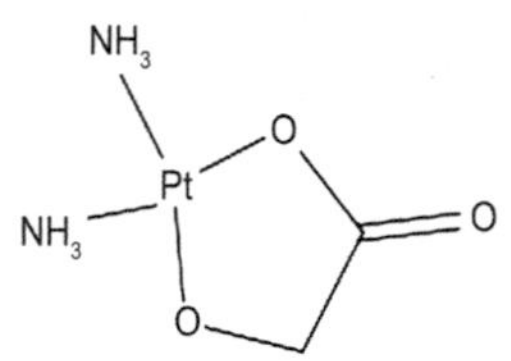

Figure 2.23. Carboplatin (Diamine cyclobutane dicarboxylate platinum(II)); cis-diamine-1,1-cy clobutanedicarboxylatoplatinum(II);cis-diamine-(cyclobutane-1,1-dicarboxylato)-platinum(II).

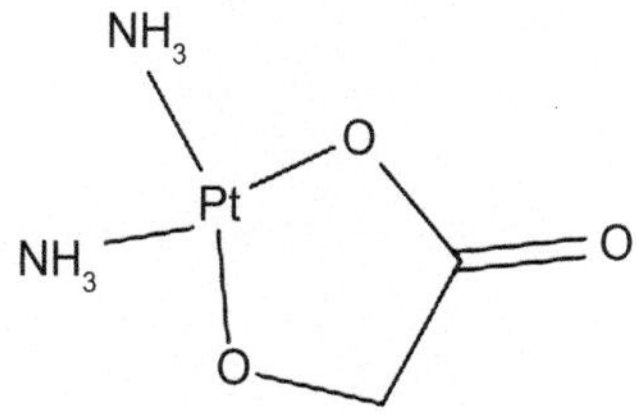

Figure 2.24. Nedaplatin (cis-diamineglycolatoplatinum (II)).

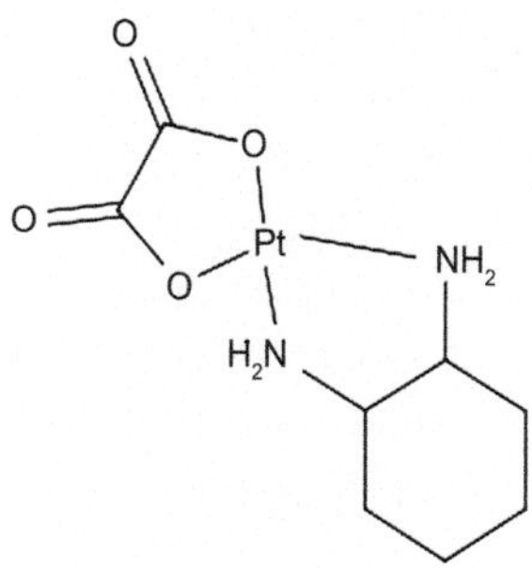

Figure 2.25. Oxaliplatin (Oxalatoplatinum(II)).

2.4. ACTIVE FORM OF CISPLATIN

Whereas the active forms of polymeric derivatives of cisplatin can vary, it may well be helpful to understand how well the active component of cisplatin itself is active both against cancers and viruses as a preliminary step for comprehending how polymeric forms are effective against both cancers and viruses.

Cisplatin is observed to undergo spontaneous hydrolysis in an aqueous solution. As indicated in Eq. 2.1, the reaction creates species such as monoaquo platinum and diaqua platinum complexes as a result of nucleophilic substitution in water.

```
H3N Cl          H3N Cl           H3N   OH2 +1
  \ /             \ /              \ /
   Pt + H2O  —>    Pt  +  H2O  —>   Pt
  /\              / \              / \
H3N Cl         H3N  OH2 +1      H3N   OH2 +1
```

Eq. 2.1

Moreover, Other aqueous species can occur, including hydroxy complexes, as shown in Figures 2.26. and 2.27. The hydroxyl species is pH dependent. At pH 7.4, 85 percent of the monohydrated complex is in the less reactive dihydroxy state. When the pH is reduced to 6.0, the most frequent form (80 percent) is the monohydrate species.

As a result, the number of potential aquated forms formed from cisplatin is large and highly dependent on pH, temperature, duration, and the concentration of related reactants such as chloride ion and ammonia. Diagram 2.1 has structures of several of these cisplatin aquated forms, including those already mentioned.

The relatively high chloride concentration (about 100 mM) in blood minimizes hydrolysis and the formation of aquated species. Once inside of the cell, where the chloride ion concentration is much lower (~4 mM), hydrolysis readily occurs giving a number of aquated species including the diaqua complex. At 37 °C the half-life for the completion of the formation of the diaqua complex is 1.7 hours with an activation energy of about 20 Kcal/mol (80 KJ/mol). Although the active form within the cell is believed to be the monohydrated structure (Figure 2.28.), the "preferred" extracellular species contains two cis-oriented leaving groups that are normally chloride ligands.

As noted, before, due to the high chloride ion concentration in blood these leaving groups will remain in position resulting in the molecule being electrically neutral until it enters the cell, as a result, the molecule is electrically neutral until it reaches the cell.

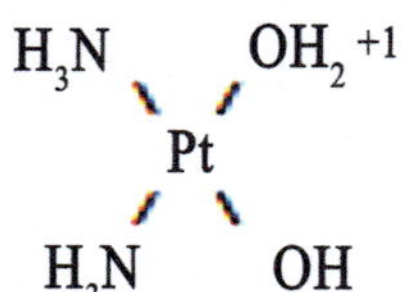

26

Figure 2.26. Selected aquated forms of cisplatin.

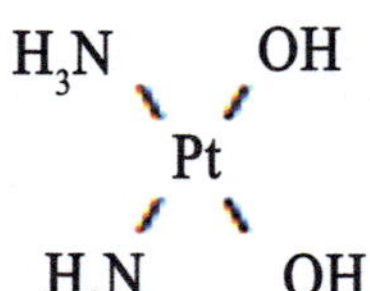

27

Figure 2.27. Selected aquated forms of cisplatin.

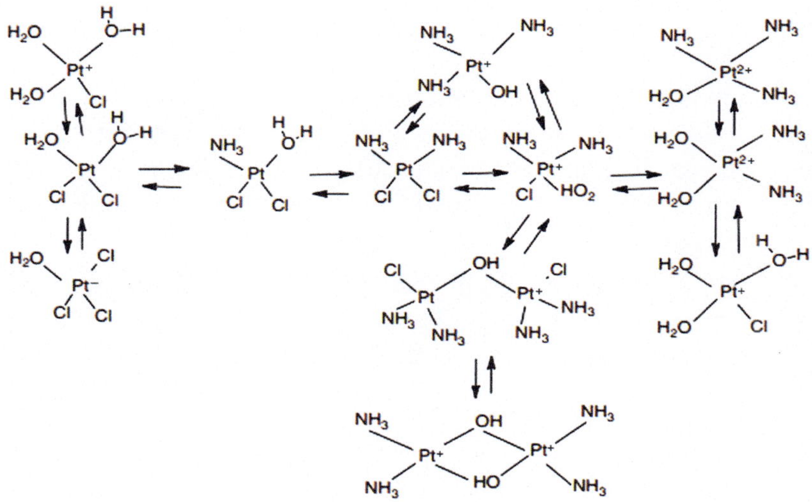

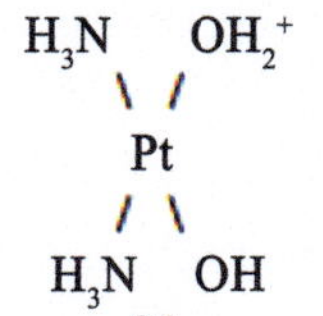

28

Figure 2.28. Selected aquated forms of cisplatin.

As previously stated, cis-DDP reaches cells through diffusion and is transformed into an active form. The decreased intracellular chloride concentration increases ligand exchange of chloride for water, resulting in the formation of the active aquated complex. To enter the cell, the platinum-containing complex must be neutral, and labile chloride groups must be present to generate the active species within the cell.

As a monoaquo species, cis-antineoplastic DDP's effect appears to be due to its interaction with DNA nucleotides. The monohydrated complex combines with the nucleotides in the DNA to create intra and interstrand crosslinks. Cis-DDP has been demonstrated to preferentially interact with guanine among the four nucleic acid bases. Intrastrand crosslinks between neighboring guanines are by far the most prevalent.

There are various conceivable DNA crosslinks. In an A-T rich area, one preferred interstrand option occurs between the 6-NH groups of adenines on opposing strands. This is due to the fact that these groups are around 3.5 apart, which is close to the 3 distance between the cis leaving groups on the platinum.

Cross-linking between the amino groups guanine and cytosine in opposing strands is the second preferred approach. This is preferred because the platinum is at right angles to the bases, which are coplanar. This means that the bases must either "bend down" or "turn edge" to reach the required conformation to bind to the platinum complex.

This binding arrangement is thought to cause secondary structure disruption and mild double helix disruption. This is enough to block DNA replication and transcription, causing cell death, but too tiny to trigger a reaction by damage detection proteins, resulting in excision of the damaged section and strand repair.

Though cisplatin may generate many different forms of crosslinks, intrastrand crosslinks are the most prevalent. The majority of these crosslinks can be repaired, however, at least one form of interstrand crosslink may not cause cellular repair enzymes to respond. Furthermore, while intrastrand crosslinks are significant in characterizing cisplatin action, triggering apoptosis appears to be a role in cisplatin's anticancer process.

Whereas cis-DDP is thought to function within the cell, other platinum-containing compounds, such as the so-called "platinum blues," appear to act on the cell membrane. As a result, the specific manner and location of activity may differ. Although cisplatin's undeniable effectiveness, restrictions exist, including severe toxic side effects. These toxic side effects include gastrointestinal problems such as acute nausea, vomiting, and diarrhea; rare cases of liver dysfunction; myelosuppression involving anemia, leukopenia, and thrombocytopenia; nephrotoxicity; and, less frequently, immunosuppression, hypo magnesia, hypocalcemia, and cardiotoxicity. The most significant adverse effect is renal damage. Most of the cis-DDP supplied is filtered out of the body within a few hours, exposing the kidneys to bursts of high platinum concentrations. The renal difficulties are thought to be caused by the quick pace at which the kidneys filter platinum from the blood. Another issue is the permanent and cumulative hearing loss.

2.5. STRUCTURE-ACTIVITY RELATIONSHIPS

The issue of structure-activity correlations is difficult and poorly established for anticancer action, and its potential to suppress viruses is unclear. Most approaches involving cisplatin polymeric derivatives are predicated in part on the premise that the platinum-containing polymer will behave similarly to cisplatin itself.

However, it is probable that at least some of these polymeric medicines limit cancer development in other ways. Although this complicates overall methods, it may be a benefit since they would work in a different way to supplement cisplatin as a medicine.

As a result, they may act in a variety of ways, allowing cancer development to be inhibited by a variety of pathways. Nonetheless, it may be useful to review some structure-activity connections identified for cisplatin as an anticancer drug.

Despite the fact that numerous promising products have been developed, underlying structure-activity connections remain unknown for a variety of reasons, not the least of which is that various platinum-containing compounds may suppress cancer development in different ways. As a result, the following discussion should be seen as a single quick attempt to describe broad structure-activity correlations.

First, there should be two anionic leaving groups available, such as chloride, bromide, or oxalate. Except for the exceptionally labile malonato ligand, bidenate chelating groups such as dicarboxylate dianions are frequently preferred to monodentate ligands due to their greater ability to remain intact in the circulation.

Complexes with more labile groups, such as the nitrate ion, hydrolyze too quickly for *in vivo* application, while ligands, such as the cyanide ion, bind the platinum too firmly, reducing its activity. Furthermore, such complexes should have cis geometry and be neutral, with moderately innocuous amine or nitrogen donor groups. The molecule's neutrality is thought to help the platinum-containing medication to pass more easily across the cell membrane. The amines should really be primary or secondary amines, allowing for hydrogen bonding. Because of the comparatively high Pt–N bond strength, amines and similar ligands are tightly bound, whereas leaving groups such as chlorides and carboxylate anions are weakly bonded and easily replaced by other nucleophiles. With these aqua structures that are replaceable, a multitude of aqua complexes is produced.

Whenever the pH exceeds 6.0, chloride displacement by the hydroxyl anion is preferred, resulting in complexes containing the hydroxo group, a generally weak leaving group.

2.6. ARGUMENTS FOR CISPLATIN-DERIVATIVE DRUGS

Since Rosenberg discovered that cisplatin is an excellent anti-cancer agent, major synthesis and assessment projects have sought to develop cisplatin derivatives with stronger and more widespread anticancer efficacy but lower toxicity. Recent research has concentrated on the development of platinum-containing homing compounds that function directly at the desired tumor location.

Oncologists all around the world have used the cytotoxic activity of a wide range of medication systems to treat cancer for decades. Despite undeniable success in cancer chemotherapy, particularly in conjunction with surgery and other treatment modalities, multiple significant pharmacological shortcomings of anticancer medications have been widely acknowledged in the medical community.

Most medications lack cell specificity, making it impossible to distinguish between normal and malignant cells. As a result, they frequently produce severe and dose-limiting systemic toxicity. They instantly reveal themselves as targets for scavenger proteins or as substrates for glomerular filtration and first-pass liver metabolism as extraneous agents.

As a result, serum residence periods are frequently short, the majority of given dosages are prematurely eliminated (and hence squandered), and bioavailability (concentration in the target tissue) is typically poor. Many medications have polar, charged, or salt-like properties. As a result, they are poor substrates for membrane penetration, intracellular trafficking, and cell entrance via the passive diffusion process that is shared by neutral and nonpolar substances.

Furthermore, medicines with low water solubility are slowly and incompletely dispersed in the central circulation, making them accessible targets for the reticuloendothelial system. Finally, and most significantly, acquired drug resistance, which progressively develops in target cells after initially effective chemotherapy, is a reasonably common event that necessitates early treatment discontinuation. As a result of these flaws, the therapeutic window is restricted and overall chemotherapeutic efficacy is severely reduced.

Another method to overcome these shortcomings is to turn the active substance into a prodrug which will experience little hindrance from scavenger processes. This prodrug would be able to penetrate intercellular membranes and approach the target location, which in cancer treatment implies breaking into the malignant cell's lysosomal compartment. One such kind of prodrug is characterized by the inclusion of a platinum-containing compound in polymers.

2.7. ARGUMENTS FOR POLYMERIC DRUGS

Polymers can function as transporters, medications, or a mix of the two. This subject is covered in greater depth in research. Polymers with numerous positive components, such as homing devices and the "bullet," can be created as carriers. The polymer can also be changed to obtain solubility in water, appropriate control release kinetics, a balance of hydrophilic and hydrophobic character, as well as nontoxic and nonimmunogenic properties.

Its size serves to protect the medicine from serum protein attack, resulting in longer serum circulation half-lives. An often-mentioned molecular weight range is 25,000 to 80,000, which will delay early renal excretion while limiting harmful consequences as evidenced by high molecular weight polymers occasionally.

The carrier-attached medication, whether polar or charged, will be delivered into the intracellular space through a pinocytotic cell entry mechanism, overcoming probable influx inhibition or efflux acceleration caused by certain well-defined drug resistance mechanisms. The inclusion of potentially cationic moieties, such as tert-amino groups, speeds up the process. Cationic regions in polymers are known to improve pinocytosis while also boosting affinity for the neoplastic cell, which is negatively charged in many malignancies. Finally, the enhanced permeability and retention (EPR) effect associated with macromolecules, as opposed to tiny chemicals, allows polymers to be distributed preferentially to tumoros tissue.

As a result, conjugate buildup in the tumor is preferential to that in healthy tissue. As a result, systemic toxicity is decreased as well as increased bioavailability the drug-containing polymer can also function as a drug.

The incorporation of the cisplatin-like molecule into a polymer is expected to accomplish the following:

1. It will restrict the physiologically active drug's movement. Polymers, due to their bulk, are more difficult to penetrate through bodily membranes. Cisplatin is rapidly eliminated from the body, exposing the kidney and other organs to high levels of platinum.

Polymers having chain lengths of 100 units or more often have difficulty moving across biological membranes. Restricted mobility may reduce renal and other organ damage by preventing a buildup in the kidneys and other organs. Moreover, platinum from polymers might be released gradually, decreasing organ exposure to high concentrations of platinum-containing complexes.

2. This may boost activity by increasing the possibility of numerous bonding contacts at a given location (e.g., chemical bonding, hydrogen bonding, hydrophobic interactions).
3. It should improve bioactive moiety delivery while decreasing toxicity. Cisplatin hydrolyses in aqueous solutions with a reaction half-life of nine hours at ambient temperature or 2.4 hours at 37 °C. Cisplatin hydrolyzes in the body, producing a wide range of platinum-containing compounds, none of which are as active as cisplatin and most of which are hazardous to the body.

The formation of these hydrolysis products increases the quantity of platinum complex required to achieve the desired tumors decrease. As a result, the number of platinum complexes that must be digested by the body rises. The polymeric structure should also protect the platinum moiety from undesired hydrolysis, so by increasing the concentration of platinum in the favorable form preserved in the body as a result, smaller effective dosages of the medicine can be utilized. The more hydrophobic polymer chain should also shield the platinum moiety from direct water attack.

4. It should be able to avoid the cell's protection mechanism. As a result of the invasion of other chemo medicines, the cell's defense reaction is active. According to recent research, the entry of chemotherapy medications into cells induces the accumulation of "housekeeping" proteins, which are fairly universal in their capacity to pick and eliminate foreign chemicals existing in the cell. This might be one of the main reasons why chemo treatments cause drug resistance, even to medications that have never been used before. The polymeric structure of the platinum carriers may deter housekeeping proteins from removing them, allowing the polymers to operate as anticancer medications in situations where smaller platinum-containing pharmaceuticals fail.

2.8. POLYMER SYNTHESIS

The polymers were made by combining equal molar water solutions of potassium tetrachloroplatinate II and the diamine-containing reactant, and the resultant polymer was collected as a precipitate. Polymer formation can take anywhere from a few minutes to many days, depending on the reactivity of the Lewis base.

Polymers can be produced in gram amounts or in greater quantities as needed, using basic equipment and commercially accessible reactants. As a result, the polymers are well suited for commercialization.

2.9. ANTIVIRAL ACTIVITY

Certain malignancies are thought to have a viral connection. As such, studying the viral response to certain platinum polymers is instructive. Tetramisole is an antihelmintic that works by inhibiting cyclic nucleotide phosphodiesterases. It is actually a mixture of optical isomers, the most active of which is levamisole. Levamisole was the first synthetic molecule to be discovered to have immunomodulatory effects. It seems to return macrophage and T-lymphocyte functioning to normal.

Cisplatin polymer analogues created by reacting tetrachloroplatinate with tetramisole were investigated in ICR Swiss male mice for their potential to suppress EMC-D viruses that cause the emergence of juvenile diabetes symptoms.

In brief, the mice were given doses of 1, 5, and 10 mg/kg. In comparison to untreated mice, doses of 1 and 5 mg/kg reduced the severity and incidence of virus-induced diabetes. In another set of testing, dosages of 1 and 10 mg/kg were given 1 day before the viral injection, however, there was an increase in the severity and incidence of virus-induced diabetes. Other tests revealed that the polymer had distinct activity profiles than the tetramisole (Figure 2.29.).

Methotrexate is a folic acid antagonist that inhibits purine synthesis indirectly. It is especially effective in rapidly reproducing cell types such as cancer. It inhibits the production of primary and secondary antibodies, the homograft reaction, the graft-versus-host response, and the development of hypersensitivity. Methotrexate is used to cure some malignancies. Long-term usage is extremely harmful, especially to the liver. The usage of folinic acid reverses the harmful impact of methotrexate. This "discovery" resulted in the invention of the "rescue procedure," in which the most quickly proliferating cells are destroyed while the remaining cells remain unaffected.

The polymers were created by combining equal molar water solutions of potassium tetrachloroplatinate II and methotrexate, and the resultant polymer

(Figure 2.30.) was collected as a precipitate. A methotrexate polymer (Figure 2.30.) was evaluated in research comparable to that of tetramisole. Female mice were treated in the original trial. Only male IRC Swiss mice acquire diabetes-like symptoms. Female mice must first be treated with testosterone before they can acquire diabetes-like symptoms.

The mice were placed into three groups, each of which got polymer injections (0.5 cc IP of a polymer solution containing 6.4 mg/kg). A week later, testosterone was administered to Groups I and II. On day 8, group 1 received a second 0.5 cc intraperitoneal (i.p.) inoculation of the polymer solution. On day 9, all groups got 1 104 pfu (plaque forming units) of EMC-D virus. On day 17, all mice were subjected to a one-hour glucose tolerance test.

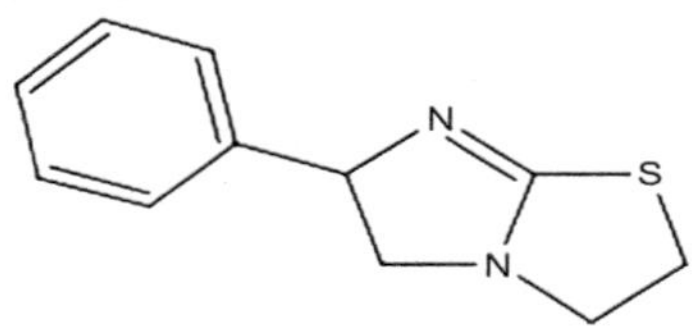

Figure 2.29. Tetramisole.

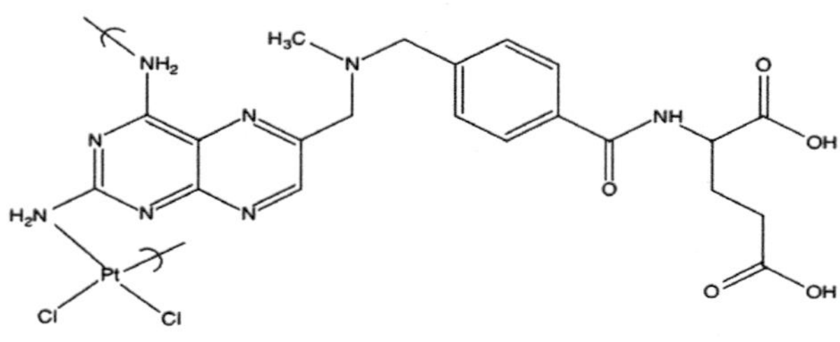

Figure 2.30. Methotrexate Polymer.

The glucose levels in groups I and III were comparable and much lower than the levels in diabetic mice in group II. This is consistent with the polymer successfully inhibiting the virus's diabetogenic effects. Furthermore, other findings from this study suggested that this breed of female mice was vulnerable to developing diabetes-like symptoms even without testosterone therapy.

A similar investigation, but with male mice, was conducted. Again, the polymer (Figure 2.30.) had a stronger favorable effect on diabetes management than either of the reactants. For the polymer-treated animals, glucose levels were comparable to noninfected mice.

Again, the integration of both platinum and methotrexate into a polymer proved an effective agent, rather than either medication alone. These two investigations are connected to the development of a vaccination that can be used to prevent the onset of -cell damage caused by RNA viruses.

The third trial concentrated on therapy after viral infection. The polymer was 100 percent efficient in viral control when administered 1 day after the mice were infected (Figure 2.30.). In conclusion, the methotrexate polymer (Figure 2.30.) is a potent antiviral agent against at least the EMC RNA virus.

Researchers recently investigated the antiviral activity of methotrexate, tetrachloroplatinate, a physical combination of methotrexate and tetrachloroplatinate, and a methotrexate-platinum polymer. A similar investigation was conducted using tilorone and a tilorone derivative, as well as cisplatin polymeric compounds.

For these research assays, each cell line was specifically chosen to be compatible with the virus's development. DSC-1 cells are kidney epithelial cells from African green monkeys, mouse L929 cells are fibroblast cells, Vero cells are kidney epithelial cells from African green monkeys, and human 143 cells are fibroblast bone osteosarcoma cells.

The viruses were chosen to illustrate a wide spectrum of viruses. The reovirus ST3 virus is an RNA virus that is now being studied due to its potential to suppress particular cancer cells while leaving normal cells alone. In general, medications that can block one RNA virus will be effective against other RNA viruses.

The other viruses are all DNA viruses, and their action against distinct DNA viruses must be researched separately. Vaccinia causes smallpox; herpes simplex causes at least 45 million infections every year in the United States, or one out of every five adolescents and adults; and varicella-zoster causes chickenpox and shingles.

The cell lines utilized for viral replication experiments are cancer cell lines in order to provide a measure of the test agents' capacity to limit cell proliferation. Each cell line is carefully selected to be compatible with the virus's proliferation. All of the cell lines have been transformed. Table 2.1 shows the GI50 values for these chemicals in g/ml for the various cell lines. Viral replication experiments are conducted at doses where cell mortality is less than 5%.

For comparison, Cisplatin has a GI50 of 50 g/ml in L929 cells. Therefore, with the exception of tetrachloroplatinate, all of the examined compounds have GI50 values lower than cisplatin and, with the exception of tetrachloroplatinate, equal ability to suppress cell growth.

Table 2.1.Toxicity of methotrexate-related compounds to cancer cell lines

Compound	GI_{50}(microgram/mL)			
	L929 Cells	143 Cells	Vero Cells	BS-C-1 Cell
K_2PtCl_4	375	275	225	225
Methotrexate	15	10	12	10
Polymer	10	10	10	12
Methotrexate Mix	12	8	10	10

Table 2.2. Inhibition concentrations **(μg/ml) for the tested compounds**

→Tested compound	K_2PtCl_4	Methotrexate	Polymer mixture	
Reovirus ST3				
GI50	-			
GI100	-			
*GI150	-			
*GI100	-			
Vacinia WR				
GI50	-			
GI100	-		8	
*GI150	-		4	
*GI100	-			
HSV-I				
GI50	-	2	3	2
GI100	-	8	6	4
*GI150	-	2	3	2
*GI100	-	8	3	2
VZV				
GI50	-	3	4	4
GI100	-	8	4	6
*GI150	-	3	2	2
*GI100	-	8	4	3

Tables 2.1 and 2.2 present early findings on the potential of several polymeric medicines to suppress DNA and RNA related viruses. The

findings are provided in the form of four experiments, each with duplicate samples. Table 2.2 summarizes the capacity of each chemical to suppress viral growth using a plaque reduction test. Tetrachlorate is virtually inert against the viruses examined. Methotrexate, tetrachloroplatinate, and the polymer all have good activity against HSV-1 and VZV, both DNA viruses whose genome replication occurs in the nucleus, with some activity against vaccinia virus, a DNA virus with cytoplasmic DNA replication, and no activity against reovirus, a dsRNA virus with cytoplasmic replication.

At lower doses than methotrexate, the polymer and physical combination demonstrate excellent action. Table 2.3 also includes the GI50 and GI100 values for the different compounds, as well as columns depending on the quantity of methotrexate itself, because much of the action appears to be due to the presence of methotrexate and little, if any, from tetrachloroplatinate.

It is noticed that the polymer often has the lowest inhibition concentrations, which is consistent with the polymer's stronger capacity to block the tested viruses. Polymer activity may occur from methotrexate-controlled release, the polymer itself functioning as a medication, or some mix of the two extremes.

In conclusion, methotrexate, the combination, and the polymer are all active against HSV-1 and VZV viruses, however, tetrachloroplatinate had little or no action against any of the viruses tested. The polymer and the combination showed the best inhibition, that is, inhibition at the lowest concentration, consistent with a cooperative impact of the methotrexate and platinum moieties.

It is probable that when tetrachloroplatinate and methotrexate were combined together in solution, they created a polymer, which is consistent with comparable findings for the combination and polymer.

The same research was done using tilorone and a tilorone derivative (Figure 2.31.). A molecular compound comprising tilorone and RNA is known to have an antiviral action comparable to polynucleotide interferon (IFN) inducers such as poly(I)-poly(C), larifan, and ridostin. It's probable that the present reactions are connected to this.

Tilorone, 2,7-bis[2-9diethylamino) ethoxy]-9H-fluorene-9-one is the first synthesized small molecule structure that has been identified as an orally active interferon inducer. A number of comparable structures were synthesised due to their potential relevance. These derivatives are assigned different numerals after the term tilorone. One of these derivatives is tilorone.

Figures 2.32. and 2.33. show the structures of platinum polymers. Each is a cis-derivative that follows the trans-effect. Table 2.3. shows the toxicity of tilorone polymers to several cell lines in ng/ml. Both chemicals had equivalent toxicity to the cancer cell lines studied, but not to the partly transformed Balb 3T3 cells.

To compare, cisplatin has a GI50 of 50,000 ng/ml, which is more than 250 times that of any platinum-tilorone polymer. Table 2.4.summarizes preliminary research on the efficacy of Pt-tilorone polymeric medicines to inhibit DNA and RNA related viruses.

Table 2.3. Toxicity of tillorone compounds to various cell lines

Compound	GI_{50}(nanogram/mL)				
	L929 Cells	143 Cells	Vero Cells	BS-C-1 Cells	Balb 3T3 Cells
Pt-Tilorone 11.567	225	200	175	200	10,000
Pt-Tilorone	180	125	175	200	500

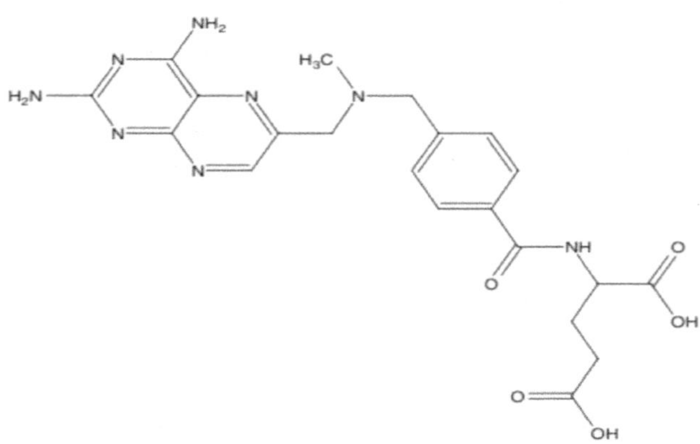

Figure 2.31. Tilorone.

At small doses, both of the tillorone polymers inhibited all of the viruses tested. In comparison, the organotin polymers of norfloxacin and ampicillin inhibited these identical viruses effectively at somewhat greater doses,

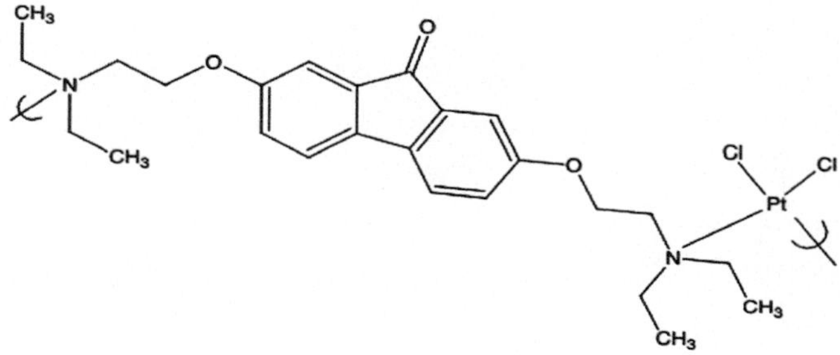

Figure 2.32. Product of potassium tetrachloroplatinate II and tilorone.

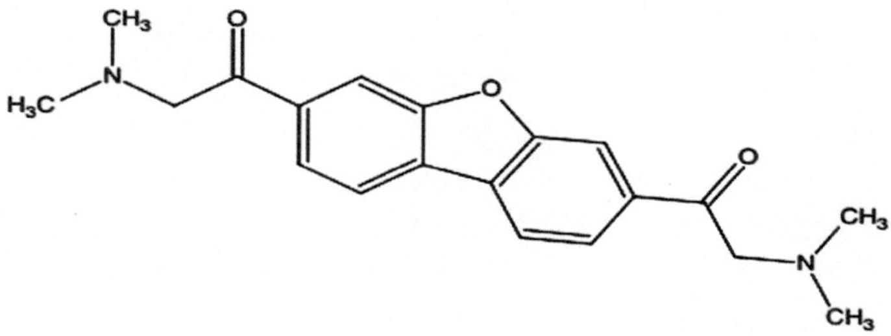

Figure 2.33. Product of potassium tetrachloroplatinate II and tilorone 11,567.

Table 2.4. Plaque-reduction assay results for the platinum-tilorone polymers

Compound	GI_{99+} Concentration(ng/ml)			
	Reovirus ST3	Vaccinia WP	HSV-I	BS-C-1 C
Pt-Tilorone 11.567	200	150	150	150
Pt-Tilorone	125	100	125	100

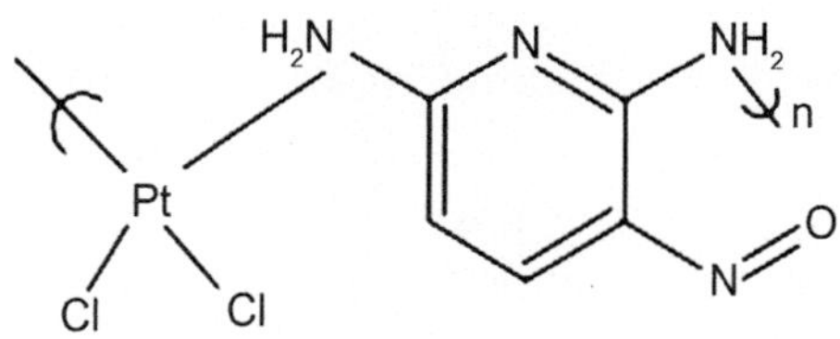

Figure 2.34. Polymer derived from tetrachloroplatinate.

within in the range of less than 1 to 2 mg/ml, whereas tilorone compounds demonstrated excellent efficacy at 0.2 mg/ml, as previously stated. To summarize, tilorone polymers block both RNA and DNA viruses and should be investigated further as an antiviral drug in the fight against viruses and potential bioterrorism employing viruses.

Furthermore, they demonstrate excellent reduction of virus replication in both transformed and normal cell lines, a state that more closely resembles human antiviral treatment.

A number of platinum polyamines have been evaluated in tumor cells for antiviral efficacy. For example, a polymer derived from tetrachloroplatinate and 2,6-diamino3-nitroso-pyridine (Figure 2.34.) with a cell differential ratio of 3.4 was evaluated on L929 cells infected with Encephalomyocarditis at a dosage of 2.2 g/ml (EMC)MM virus strain A viral decrease of around 25% was seen. This is regarded as a moderate antiviral reaction.

Agents generally, that inhibit one RNA virus will often inhibit other RNA viruses, whereas each DNA virus must be studied independently. Platinum polyamines were tested for their antiviral activity against RNA viruses. The behavior toward RNA viruses varied, with some displaying minimal action but the majority indicating suppression of viral reproduction at polymer concentrations lower than that required for tumor inhibition (1 g/ml).

The effect of platinum polyamines on SV40 viral transformation of 3T3 cells was also investigated. In conclusion, these polymers had no influence on the process of transformation.

2.10. VANADOCENE-CONTAINING POLYMERS

Mostly all vanadium-containing organometallics are generated primarily from dicyclopentanylenevanadium (IV) dichloride and vanadocene dichloride. Vanadocene dichloride and its derivatives belong to a class of chemicals known as deformed sandwich compounds. Vanadocene dichloride, unlike ferrocene, has distorted tetrahedral geometry, with the cyclopentadienes facing the vanadium atom. It has structural similarities to a variety of other metallocene dichlorides, including those with Ti, Zr, Hf, Mo, and Nb as the metal. These compounds have comparable geometries as well as chemical reactions and biological activity.

The primary application of these metallocenes is as catalysts in the manufacture of a variety of stereoregular vinyl polymers such as polyethylene and polypropylene. These are also being studied for their biological properties.

Because the biological activity is a primary driving force in the synthesis of polymeric derivatives, we will quickly cover some of this work, concentrating on vanadocene dichloride as the molecule more commonly researched.

2.11. ANTICANCER ACTIVITY

The deformed tetrahedral metallocene dihalides are a type of hydrophobic organometallic anticancer drug that inhibits several cancer cell types. They inhibit leukemias P388, L1210, colon 38, Lewis lung carcinomas, B16 melanoma, and solid and fluid Ehrilich ascites tumors. Titanocene dichloride, which is now in phase II clinical studies, is active against xenografted malignancies of the rectum, stomach, colon, breast, neck, and stomach.

Titanocene and vanadocene dichloride both impede DNA and RNA replication in nucleic acid-rich tumor cells. As a result, it is thought that the formation of DNA-metallocene adducts *in vivo* plays a role in their capacity to suppress cell proliferation. Furthermore, metallocene-protein interactions have been discovered to be crucial, and metal-tranferrin compounds have been incorporated into their mechanism of action.

Proteins involved in cellular replication, in addition to blood plasma proteins, have been linked to the activity of many metallocenes. Vanadocene dichloride inhibits the activities of both protein kinase C and bacterial topoisomerase. Mokdsi and Harding () recently investigated the suppression of human topoisomerase II using a variety of metallocenes.

Their findings suggest that the mechanism of action includes a complicated route involving many species that modulates the transport and delivery of active chemicals to cancer cells. There is also a subsequent interaction with nucleic acids and/or proteins. While toposiomerase II inhibition is a plausible site of action, additional sites are not ruled out. Several recent reviews in this field have been published. Unlike cisplatin, vanadocene complexes did not stimulate DEL recombination in yeast or any of the DNA damage related promoters in HepG2 cells. Vanadocene compounds activate the c-fos promoter while having no effect on the minimum promoter including p53 response elements or the GADD45 promoter.

The findings supported the idea that the apoptotic signal associated with the presence of vanadocene chemicals is not generated by primary DNA damage and does not need p53 activation. As a result, it appears to regulate cell development differently than cisplatin and may be used in addition to it.

It should be noted that many metallocene derivatives lack the previously mentioned biological activities, hence their existence does not automatically confer biological activity on molecules that contain this moiety.

2.12. SPERMICIDAL ACTIVITY

The Wayne Hughes Institute has pioneered studies on the use of metallocene complexes as spermicidal medicines, with an emphasis on vanadocene derivatives. Several vanadium salts were discovered to operate as cellular redox potential modulators and to have pleiotropic effects in a variety of biological systems by accelerating the formation of active oxygen molecules.

These reactive oxygen species influence sperm motility by per oxidizing membrane lipids and proteins. Peroxidative damage to the sperm plasma membranes is also thought to have a role in the development of male infertility. Because of the large amount of unsaturated fatty acids in sperm and the relative scarcity of cytoplasmic enzymes that scavenge reactive oxygen molecules that trigger lipid peroxidation, sperm are vulnerable to oxidative stress.

Researchers investigated metallocenes including Ti, Zr, Mo, and Hf and discovered that they lacked spermicidal action. In comparison, a large variety of vanadocene molecules have spermicidal action.

In cancer cells, vanadocene dichloride impairs mitotic spindle formation. Although the control cancer cells had organized mitotic spindles organized as a bipolar microtubule array with the DNA organized on a metaphase plate, the vanadocene dichloride treated cancer cells had aberrant monopolar mitotic structures with microtubules on only one side of the chromosomes arranged in a circular pattern. This demonstrates that vanadocene dichloride can inhibit cell division by interfering with bipolar spindle formation, resulting in a halt in the G2/M phase of cell cycle reproduction. A consequence of the research into vanadocenes as antiproliferative chemicals has shown a new pathway for their anticancer action.

2.13. FIBERS

In a separate investigation, it was discovered that when vanadocene polymers are separated, they form fibers. Nonfibrous flat plate-like objects were intermingled in with the fibers. Fibers have been discovered for a variety of metal-containing condensation polymers, typically those with stiff backbones. The hard guanosine part of the acyclovir backbone is joined by the more flexible ether-tail.

The existence of the guanosine ring system appears to be sufficient to trigger fiber production. Some of the threads were so tiny that they were not visible to the naked eye and could only be seen with tenfold magnification. There seemed to be two kinds of fibers: one which was relatively short (with lengths of around 0.4 mm) and one which is thicker (diameter of about 0.01 mm) with aspect ratios ranging from 40 to 100. These fibers looked to have multiple spurs going off the primary shoot and were exceedingly plentiful.

The second fibers were significantly longer (4 mm) and much thinner (0.01 mm), with aspect ratios ranging from 400 to 1,000. They were more smooth, flexible, and transparent. As previously stated, this was not the only report of such unusual fibers. Mechanical testing should be performed on these fibers, and their potential in composites should be identified.

2.14. EXPERIMENTAL: SYNTHESIS AND PHYSICAL CHARACTERIZATION

The reactants were used exactly as received. Sigma, St. Louis, provided tilorone and tilorone 11,567, while J & J Materials, Neptune City, NJ provided potassium tetrachloroplatinate(II). Synthesis was performed in the normal method. At room temperature, tilorone (5.4 mmole) dissolved in 150 mL distilled water was added to stirred solutions of potassium tetrachloroplatinate(II) (5.4 mmole) in 15 mL distilled water.

After roughly an hour, product precipitation began. After 6 hours, the product (in better than 90% yield) was collected by vacuum filtering and rinsed twice with distilled water to eliminate unreacted impurities. The solid was rinsed and dried in a glass petri dish.

The molecular weight was measured using a Brice-Phoenix BP-3000 Universal Light Scattering Photometer. A Bausch Lomb Abbe Model 3-L refractometer was used to measure the refractive indices in DMSO. By dissolving 1–10 mg of polymer in 3 mL of liquid, solubilities were measured. The solid-liquid mixtures were studied for 2–4 weeks.

KBr pellets were used to generate infrared spectra on a Mattson Instruments galaxy Series 4020 FTIR with 32 scans and an instrumental resolution of 4 1/cm. Varian-500 and Varian-400 spectrometers were used to gather proton NMR data using DMSOd6 and D2O.

Two methodologies were used for mass spectral analysis. A direct insertion probe linked to a Kratos MS-50 mass spectrophotometer operating in EI mode, 8 kV acceleration, and 10 s/decade scan rate with a probe temperature of 350–450 °C was used for the HR MS study. A Voyager-DE STR Bio Spectrometer, Applied Biosystems, Foster City, CA, was used for high-resolution electron impact positive ion matrix aided laser desorption ionization time of flight, HR MALDI-TOF, mass spectrometry.

Standard settings were employed with a linear mode of operation, an accelerating voltage of 25,000 volts, a grid voltage of 90%, and an acquisition mass range of 2,000–100,000 gram. Usually, two hundred images were collected for each spectrum. Several matrix materials were used, however, only findings using 2,5-dihydroxybenzoic acid are given in this publication.

2.15. EXPERIMENTAL: BIOLOGICAL CHARACTERIZATION

Cell Lines and Viruses Reovirus serotype 3 (ST3) strain Dearing was propagated in mouse L929 fiber blasts (ATCC CCL-1), Vaccinia virus, strain WR was propagated in human 143 cells, Herpes simplex virus 1, strain GHSV-UL46 (HSV-1 ATCC VR-1544) was propagated in Vero cells, and Varicella Zoster virus, strain Ellen (VSV ATCC VR-137) was propagated in BS-C-1 cells. Cell lines were grown in monolayer cultures in minimum essential medium (MEM) with Earles' salts and 5% fetal bovine serum (FBS). Cells were passaged at 1:2–1:10 dilutions using 0.05 percent trypsin and 0.02 percent EDTA, as per standard methods.

The medicines were made by dissolving the dry material in 100 percent DMSO at a concentration of 10 microgram/mL. Working stocks were created by diluting stocks 1–10 into MEM, resulting in a final drug concentration of 1 microgram/mL. The material was transferred from these working stocks to MEM with 5% FBS, providing the stated final concentrations, and the medium was applied to the cell monolayers. Cytotoxicity Tests: Drug cytotoxicity was assessed by plating cells at a concentration of 5 105 cells per well in MEM with 5% FBS in a 6-well plate and incubating the plates at 37 °C, 5% CO2 for about 24 hours until the cells split to yield 1 106 cells per well. At this point, the medium was withdrawn and replaced with MEM containing 5% FBS and the specified medication concentration. Cytotoxicity was assessed microscopically after 48 and 96 hours with the addition of trypan blue to label nonviable cells. The tests were carried out in pairs. Assays for Reovirus, Vaccinia, HSV-1 and VZV Plaque Reduction

L929, human 143, vero, or BSC-1 cells were grown to confluency in 6-well plates in MEM containing 5% FBS. In 250 microliters of MEM, cells were infected with reovirus, vaccinia, HSV-1, or VZV virus at successive 10-fold dilutions ranging from 1 106 to 10 plaque-forming units (PFUs).

After 30 minutes, the medium was changed with MEM containing 5% FBS and the specified medication concentration. Assays for Reovirus, Vaccinia, HSV-1 and VZV Plaque Reduction L929, human 143, vero, or BSC-1 cells were grown to confluency in 6-well plates in MEM containing 5% FBS. In 250 microliters of MEM, cells were infected with reovirus, vaccinia, HSV-1, or VZV virus at successive 10-fold dilutions ranging from 1 106 to 10 plaque-forming units (PFUs). After 30 minutes, the medium was changed with MEM containing 5% FBS and the specified medication concentration.

2.16. CONCLUSION

Polymeric drugs offer the opportunity to avoid some of these effects. These leaky vasculatures and limited lymphatic drainage, are typical of tumors and missing in normal tissue, resulting in the accumulation of macromolecules such as polymeric drugs in the interstitial space of a large variety of tumors. Again, the organotin polymers showed inhibition of the viruses at lower concentrations than those found for acyclovir itself. Trapped polymeric drugs can then act over a longer time as polymeric drugs themselves or in controlled release of the active drug. Polymeric drugs can act either as a drug or as a means to release the drug over a period of time.

We also looked at organotin products derived from the known antiviral drug acyclovir. Polymeric drugs offer many potential advantages over monomeric or small molecule drugs. Polymeric drugs offer a number of possible advantages over the currently employed small molecule drugs. Further, these organotin polymers are much less toxic than cisplatin, the most widely used anticancer drug. The organotin polymers from norfloxacin and ampicillin showed total inhibition to virus growth at concentrations of about 2 mg/ml, whereas norfloxacin and ampicillin, themselves, exhibited no viral inhibition. A number of polymeric derivatives of cisplatin as anticancer drugs for about 30 years has been studied. Again, synthesized polymeric drugs that inhibit cancer cell growth within the same concentration range as cisplatin itself have been made and these drugs are again, much less toxic. A number of metal-containing polymers as potential antiviral agents emphasizing both platinum and organotin-containing polymers are being explored.

REFERENCES

1. Bhagdev, K. and Sarkar, S., 2021. Benzothiazole Moiety and Its Derivatives as Antiviral Agents. *The 1st International Electronic Conference on Molecular Sciences: Druggable Targets of Emerging Infectious Diseases*, [online] Available at: <https://www.mdpi.com/2673-9992/7/1/9> [Accessed 30 June 2022].
2. Engel, J., Schonenberger, H., Lux, F. and Hilgard, P., 1987. Estrophilic cisplatin derivatives. *Cancer Treatment Reviews*, [online] 14(3-4), pp.275-283. Available at: <https://www.cancertreatmentreviews.com/article/0305-7372(87)90018-1/pdf> [Accessed 30 June 2022].
3. Roner, M. and Carraher, C., 2008. Cisplatin Derivatives as Antiviral Agents. *Inorganic and Organometallic Macromolecules*, [online] pp.193-223. Available at: <https://link.springer.com/chapter/10.1007/978-0-387-72947-3_8> [Accessed 30 June 2022].
4. Roner, M. and Carraher, C., 2008. Cisplatin Derivatives as Antiviral Agents. *Inorganic and Organometallic Macromolecules*, [online] pp.193-223. Available at: <https://www.semanticscholar.org/paper/Cisplatin-Derivatives-as-Antiviral-Agents-Roner-Carraher/510bb350cd521e7f8b5f113356cd53aeddcd2716> [Accessed 30 June 2022].
5. Roner, M., Carraher, C., Dhanji, S. and Barot, G., 2008. Antiviral and Anticancer Activity of Cisplatin Derivatives of Tilorone. *Journal of Inorganic and Organometallic Polymers and Materials*, [online] 18(3). Available at: <https://www.ncbi.nlm.nih.gov/pmc/articles/PMC7088078/#:~:text=The%20polymeric%20cisplatin%20derivatives%20prevent,in%20the%20micrograms%2FmL%20range.> [Accessed 30 June 2022].

CHAPTER 3

Macromolecules Structure and Function

CONTENTS

Considering the vast complexity of life on Earth, we could anticipate a tremendous diversity of molecules in organisms. Surprisingly, the crucially significant big molecules of all living things—from bacteria to elephants—are divided into only four categories: carbohydrates, lipids, proteins, and nucleic acids.

Members of three of these classes—carbohydrates, proteins, and nucleic acids—are massive on the molecular scale and are so referred to as macromolecules.

3.1. THE MOLECULES OF LIFE

A protein, for example, may be made up of thousands of atoms that combine to produce a molecular behemoth with a mass considerably in excess of 100,000 Daltons. Given the size and complexity of macromolecules, biochemists have established the precise structure of a large number of them. The architecture of a major biological molecule explains how it operates. Large biological molecules, including water and simple organic compounds, display distinctive emergent features due to the ordered arrangement of their atoms. In this chapter, we'll look at how macromolecules are made. The structure and function of all four kinds of big biological molecules will next be investigated: carbohydrates, lipids, proteins, and nucleic acids.

3.2. MACROMOLECULES ARE POLYMERS, BUILT FROM MONOMERS

The macromolecules in three of the four types of organic chemicals found in life—carbohydrates, proteins, and nucleic acids—are polymers, which are chain-like molecules (from the Greek polys, many, and meros, part). A polymer is a lengthy molecule made up of numerous similar or identical building units joined together by covalent bonds, similar to how a railway is made up of a chain of cars. The repeating units that act as polymer building blocks are smaller molecules known as monomers (from the Greek monos, single). Some of the molecules that serve as monomers also serve additional purposes.

3.3. THE SYNTHESIS AND BREAKDOWN OF POLYMERS

Even though each kind of polymer is composed of a distinct type of monomer, the chemical mechanisms by which cells produce and degrade polymers are essentially the same in all situations. Enzymes, specialized macromolecules that speed up chemical reactions, aid in these activities in cells. Monomers are linked through a dehydration process, which occurs when two molecules are covalently attached to each other and one water molecule is lost.

When two monomers establish a link, each monomer contributes a portion of the water molecule that is released during the reaction: A hydroxyl group (—OH) is provided by one monomer, whereas the other supplies hydrogen (—H). This process is continued as monomers are added to the chain one by one, resulting in the formation of a polymer.

Hydrolysis, which is effectively the opposite of dehydration, is used to break down polymers into monomers. The term hydrolysis refers to the process of breaking down using water (from the Greek hydro, water, and lysis, break).

The addition of a water molecule breaks the connection between the monomers, with the hydrogen from the water adhering to one monomer and the hydroxyl group bonding to the neighboring monomer. Digestion is an example of hydrolysis at action within our bodies. The majority of the organic stuff in our meals is in the form of polymers, which are much too massive to enter our bodies.

Various enzymes target the polymers in the digestive system, speeding up hydrolysis. The monomers that are released are subsequently taken into circulation and distributed to all bodily cells. These cells can then employ dehydration processes to assemble the monomers into new, distinct polymers capable of performing specialized cell tasks.

3.4. THE DIVERSITY OF POLYMERS

Every cell contains hundreds of distinct macromolecules; the collection differs from one cell type to the next, even within the same organism. Small changes in polymers, mainly DNA and proteins, account for the intrinsic variances between human siblings.

Molecular variations between unrelated individuals are higher, and those across species are much more pronounced. The living world's macromolecule diversity is tremendous, and the potential variety is virtually endless.

What is the source of such amazing diversity in life? These molecules are made up of just 40 to 50 common monomers and a few uncommon ones. Creating a wide range of polymers from such a small number of monomers is comparable to creating hundreds of thousands of sentences from only 26 letters of the alphabet.

The arrangement—the specific linear order that the components follow—is crucial. This example, however, falls well short of portraying the vast diversity of macromolecules since most biological polymers include considerably more monomers than the letters in the longest word.

Proteins, for example, are made up of 20 different types of amino acids linked together in chains hundreds of amino acids long. Life's molecular logic is basic yet elegant: Small molecules found in all organisms are organized into distinct macromolecules.

Despite this enormous variation, molecule structure and function may be generally classified. Let's take a look at each of the four primary types of big biological molecules. The huge molecules in each class exhibit emergent features that are not present in their constituent building pieces.

3.5. CARBOHYDRATES SERVE AS FUEL AND BUILDING MATERIAL

Carbohydrates comprise both sugars and sugar polymers. Monosaccharides, or simple sugars, are the simplest carbohydrates; they are the monomers from which more complex carbs are built. Disaccharides are double sugars made up of two monosaccharides linked together by a covalent bond. Carbohydrates also contain polysaccharides, which are polymers made up of numerous sugar building components.

3.5.1. Sugars

Monosaccharides (from Greek monos, single, and sacchar, sugar) contain molecular formulae that are a multiple of the unit CH_2O. The most prevalent monosaccharide, glucose ($C_6H_{12}O_6$), is critical in the chemistry of life. The hallmarks of sugar may be seen in the structure of glucose: The molecule has a carbonyl group (CO) as well as many hydroxyl groups (—OH). Sugar is either an aldose (aldehyde sugar) or a ketose depending on where the carbonyl group is located (ketone sugar). Glucose, for example, is an aldose, but fructose, a glucose isomer, is a ketose. (Most sugar names end in -ose.) Another criterion for sugar classification is the length of the carbon skeleton, which can range from three to seven carbons. Glucose, Trioses and pentoses (three and five-carbon sugars) are also frequent. Another source of variation for simple sugars is the spatial arrangement of their constituents around asymmetric carbons. (An asymmetric carbon is one that is connected to four separate atoms or groups of atoms.) For example, glucose and galactose differ solely in the arrangement of components around one asymmetric carbon. What appears to be a little distinction is big enough to give the two sugars unique forms and behavior. Although drawing glucose with a linear carbon skeleton is easy, it is not entirely realistic. Glucose molecules, like most other five- and six-carbon sugars, form rings in aqueous solutions. Monosaccharides, notable glucose, are important cellular nutrition. Cells extract energy through a sequence of processes that begin with glucose molecules in the process known as cellular respiration. Simple sugar molecules are not only a vital source of energy for cells but their carbon skeletons are also used in the formation of other tiny organic molecules such as amino acids and fatty acids.

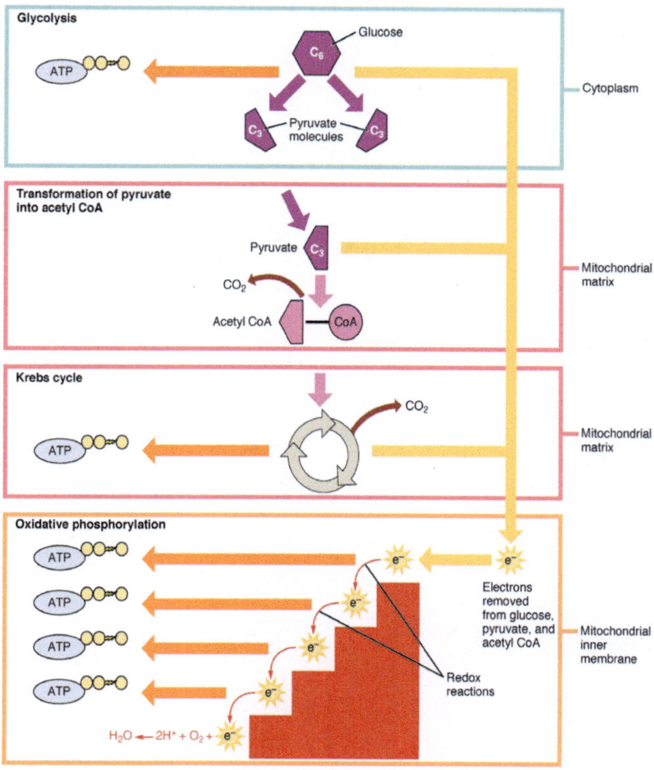

Figure 3.1. Cellular Respiration.

Source: Image by Wikimedia Commons

Sugar molecules that aren't utilized right away are usually integrated as monomers into disaccharides or polysaccharides. A disaccharide is made up of two monosaccharides connected by a glycosidic linkage, which is a covalent connection generated by a dehydration process between two monosaccharides. Maltose, for example, is a disaccharide generated by the joining of two molecules of glucose. Maltose, often known as malt sugar, is a beer-brewing component. Sucrose, or table sugar, is the most common disaccharide. It is made up of two monomers: glucose and fructose. Sucrose is commonly used by plants to carry carbohydrates from leaves to roots and other non-photosynthetic organs. Lactose, the sugar found in milk, is another disaccharide, consisting of a glucose molecule connected to a galactose molecule.

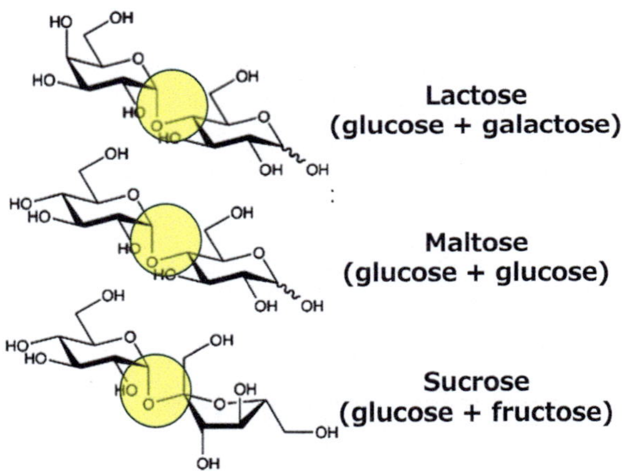

Figure 3.2. Disaccharides (Lactose, Maltose, and Sucrose).

Source: Image by Wikimedia Commons

3.5.2. Polysaccharides

Polysaccharides are macromolecules, polymers composed of hundreds to thousands of monosaccharides linked together by glycosidic connections. Some polysaccharides are used as storage materials and are hydrolyzed when needed to deliver sugar to cells.

Other polysaccharides are used to construct structures that safeguard the cell or the entire organism. A polysaccharide's architecture and function are defined by its sugar monomers and the locations of its glycosidic connections.

Storage Polysaccharides

Plants and animals both store sugars in the form of storage polysaccharides for later use. Starch, a polymer of glucose monomers, is stored in plants as granules within plastids, which contain chloroplasts.

The plant may store excess glucose by synthesizing starch. Starch indicates stored energy since glucose is a main cellular fuel. Hydrolysis, which dissolves the links between the glucose monomers, can subsequently be used to extract sugar from this carbohydrate "bank."

Most animals, including humans, contain enzymes that can hydrolyze plant starch and make glucose accessible to cells as food. The main sources of starch in the human diet are potato tubers and grains (the fruits of wheat, maize (corn), rice, and other grasses).

Like the glucose units in maltose, the majority of the glucose monomers in starch are linked by 1–4 linkages (number 1 carbon to number 4 carbons). Amylose, the most basic type of starch, is unbranched. Amylopectin is a branched polymer containing 1–6 connections at the branch sites, making it a more complicated starch.

Animals store glycogen, a polymer of glucose that is similar to amylopectin but more widely branched. Glycogen is mostly stored in the liver and muscle cells of humans and other mammals. When the need for sugar rises, glycogen hydrolysis in these cells releases glucose.

However, this stored fuel cannot maintain an animal for long. In humans, for example, glycogen reserves diminish in roughly a day unless replaced by meal consumption. This is a problem with low-carbohydrate diets.

3.5.3. Structural Polysaccharides

Structured polysaccharides are used by organisms to construct strong materials. The polysaccharide cellulose, for example, is a key component of the strong walls that surround plant cells. Plants generate over 1014 kg (100 billion tonnes) of cellulose every year on a worldwide scale; it is the most abundant organic substance on Earth.

Although cellulose, like starch, is a polymer of glucose, the glycosidic connections in these two polymers differ. The distinction is due to the fact that glucose has two slightly different ring configurations. When glucose forms a ring, the hydroxyl group connected to the number one carbon is either below or above the ring plane.

Two ring forms of glucose are known as alpha and beta. All of the glucose monomers in starch are in the confirmation. In contrast, the glucose monomers of cellulose are all in the configuration, which causes each glucose monomer to be "upside down" in relation to its neighbors.

Starch and cellulose have diverse three-dimensional forms due to their different glycosidic connections. Whereas certain starch molecules are mostly helical, cellulose molecules are straight. Cellulose is never branched, and some of the hydroxyl groups on its glucose monomers are free to hydrogen-bond with the hydroxyls of other cellulose molecules that run parallel to it. Parallel cellulose molecules linked together in this manner in plant cell walls form microfibrils. Because cellulose is the main component of paper and the single component of cotton, these cable-like microfibrils are a powerful construction material for plants as well as a vital ingredient for people.

Due to the obvious drastically different forms of these two molecules, enzymes that digest starch by hydrolyzing its links are unable to hydrolyze the connections of cellulose. Few creatures, in fact, have enzymes that can break down cellulose. Animals, including humans, do not; the cellulose in our food travels through the digestive tract and is excreted along with the excrement. The cellulose abrades the digestive system wall and induces the lining to generate mucus, which assists in the smooth flow of food through the tract. As a result, while cellulose is not a food for humans, it is a crucial component of a healthy diet.

Figure 3.3. Termites (*Nasutitermes* sp.).

Source: Image by Wikimedia Commons

The majority of fresh fruits and vegetables, as well as entire grains, are high in cellulose. "Insoluble fiber" on food containers primarily refers to cellulose. Some microbes can break down cellulose into glucose monomers. A cow's stomach contains cellulose-digesting prokaryotes and protists. These microorganisms hydrolyze the cellulose in hay and grass and convert it to glucose and other nutrients for the cow. Similarly, a termite, which cannot digest cellulose on its own, has prokaryotes or protists in its stomach that can digest wood. Certain fungi also can break down cellulose, assisting in the recycling of chemical components throughout the Earth's ecosystems.

Chitin, a carbohydrate utilized by arthropods (insects, spiders, crustaceans, and similar animals) to create their exoskeletons, is another significant structural polysaccharide. An exoskeleton is a rigid casing that protects an animal's delicate components. When coated with calcium carbonate, a salt, pure chitin turns leathery and bendable.

Chitin is also present in many fungi, which employ this polymer instead of cellulose to form their cell walls. Chitin, like cellulose, possesses connections, but the glucose monomer of chitin bears a nitrogen-containing appendage.

3.6. LIPIDS ARE A DIVERSE GROUP OF HYDROPHOBIC MOLECULES

Lipids are the only significant biological molecule class that does not include real polymers, and they are often too small to be termed macromolecules. Lipids are classified together because they have one key characteristic: they mix poorly, if at all, with water.

Their hydrophobic property is determined by their molecular structure. Although they may have some polar connections with oxygen, lipids are largely composed of hydrocarbon regions. Lipids vary in shape and function. They include waxes and some colors, but we'll concentrate on the three most essential forms of lipids: fats, phospholipids, and steroids.

3.6.1. Fats

Although fats are not polymers, they are big molecules formed by dehydration processes from smaller ones. A fat is made up of two types of smaller molecules: glycerol and fatty acids. Glycerol is alcohol with hydroxyl groups on each of its three carbons. A fatty acid has a lengthy carbon skeleton with 16 or 18 carbon atoms. The carbon at one end of the skeleton is part of a carboxyl group, which is the functional component that gives these molecules their name. The remaining skeleton is made up of a hydrocarbon chain.

Fats are hydrophobic due to the highly nonpolar CH bonds in the hydrocarbon chains of fatty acids. Fats separate from water because water molecules' hydrogen bonds with one another, excluding fats. This is why vegetable oil (a liquid fat) separates from the aqueous vinegar solution in a salad dressing container.

Three fatty acid molecules are connected to glycerol via an ester linkage, which is a connection between a hydroxyl group and a carboxyl group. The resultant fat, also known as a triacylglycerol, is made up of three fatty acids connected to a single glycerol molecule. (Another name for fat is a triglyceride, which is frequently featured in the ingredient list of packaged goods.)

The fatty acids in fat might be of the same kind or of two or three distinct types. In the context of nutrition, the phrases saturated fats and unsaturated fats are frequently employed.

These expressions relate to these phrases pertain to the structure of the fatty acid's hydrocarbon chains. If no double bonds exist between the carbon atoms in a chain, as many hydrogen atoms as feasible are bound to the carbon skeleton.

Such a structure is considered to be hydrogen-saturated, and the resultant fatty acid is known as a saturated fatty acid. An unsaturated fatty acid has one or more double bonds, with each double-bonded carbon containing one less hydrogen atom.

Almost all-natural fatty acid double bonds are cis double bonds, which generate a kink in the hydrocarbon chain wherever they occur. Saturated fat is a fat that is composed of saturated fatty acids. The majority of animal fats are saturated:

The hydrocarbon chains of their fatty acids—the "tails" of the fat molecules—lack double bonds, allowing the fat molecules to pack densely together. At room temperature, saturated animal fats such as lard and butter are solid. Plant and fish fats, on the other hand, are often unsaturated, meaning they are composed of one or more kinds of unsaturated fatty acids.

Plant and fish fats, which are usually liquid at room temperature, are referred to as oils—olive oil and cod liver oil are two examples. The bends in the cis double bonds prevent the molecules from packing tightly enough to crystallize at ambient temperature. On food labels, the phrase "hydrogenated vegetable oils" signifies that unsaturated fats have been synthetically transformed into saturated fats by the addition of hydrogen. Many goods, including peanut butter and margarine, are hydrogenated to prevent lipids from separating in liquid (oil) form.

Figure 3.4. Dalda is a brand of hydrogenated vegetable oil popular in South Asia.

Source: Image by Wikimedia Commons

A diet high in saturated fats is one of the variables that may lead to atherosclerosis, a kind of cardiovascular disease. Plaques form within the walls of blood arteries in this illness, generating inward bulges that obstruct blood flow and diminish vascular resilience.

According to recent research, the process of hydrogenating vegetable oils yields not just saturated fats but also unsaturated fats with trans double bonds. Such trans fats might cause more atherosclerosis and other issues than saturated fats.

Because Trans fats are especially prevalent in baked products and processed meals, the United States Department of Agriculture requires nutritional labels to contain trans-fat content information. Some localities in the United States, as well as at least one country (Denmark), have outlawed the use of trans fats in restaurants.

Because the body cannot produce some unsaturated fatty acids, they must be obtained from the food. Omega-3 fatty acids, which are necessary for optimal development in children and appear to protect against cardiovascular disease in adults, are among these important fatty acids.

Fatty fish, nuts, and vegetable oils are high in omega-3 fatty acids (so-called because they have a double bond at the third carbon-carbon bond from the hydrocarbon chain's terminus). Fats' primary role is energy storage. Fat hydrocarbon chains are identical to gasoline molecules and contain the same amount of energy.

A gram of fat has more than double the amount of energy as a gram of a polysaccharide like starch. Plants can function with substantial energy storage in the form of starch since they are generally immobile.

(Generally, vegetable oils are derived from seeds, where more compact storage is advantageous to the plant.)

Animals, on the other hand, must transport their energy storage, thus having a more compact reservoir of fuel—fat—has a benefit.

Other animals and humans store long-term food reserves in adipose cells, which expand and contract when fat is deposited and removed from storage. Aside from storing energy, adipose tissue also cushions critical organs like the kidneys and insulates the body with a layer of fat beneath the skin. Whales, seals, and most other marine animals have a thick subcutaneous layer that protects them from chilly ocean water.

3.6.2. Phospholipids

Phospholipids are a kind of lipid that cells cannot survive without. Phospholipids are necessary for cells because they form cell membranes. Their structure is a famous illustration of how shape and function interact at the molecular level.

A phospholipid is comparable to a fat molecule, except it only has two fatty acids bonded to glycerol instead of three. In the cell, the third hydroxyl group of glycerol is connected to a phosphate group, which has a negative electrical charge.

A variety of phospholipids may be formed by linking other tiny molecules, which are generally charged or polar, to the phosphate group. The two ends of phospholipids behave differently toward the water. Because hydrocarbon tails are hydrophobic, they are not soluble in water.

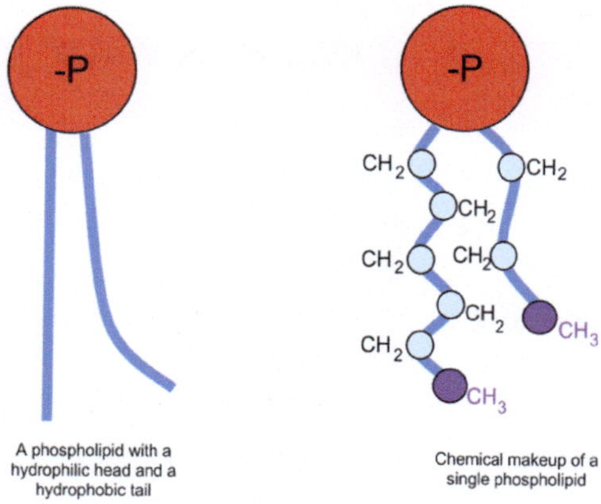

Figure 3.5. A phospholipid and the chemical makeup of the same phospholipid.

Source: Image by Wikimedia Commons

The phosphate group and its attachments, on the other hand, constitute a hydrophilic head with a high attraction for water. When phospholipids are mixed with water, they self-assemble into double-layered structures known as "bilayers," which protect the hydrophobic sections of the phospholipids from water.

Phospholipids are organized in a similar bilayer on the surface of a cell. The molecules' hydrophilic heads are on the exterior of the bilayer, in contact with the aqueous solutions within and outside the cell. The hydrophobic tails point away from the water and toward the interior of the bilayer.

The phospholipid bilayer serves as a barrier between the cell and its surroundings; in fact, cells would not survive lacking phospholipids.

3.6.3. Steroids

Steroids are lipids with a carbon skeleton made up of four fused rings. The specific chemical groups connected to this ensemble of rings identify different steroids, such as cholesterol and vertebrate sex hormones. In mammals, cholesterol is a critical chemical. It is a frequent component of animal cell membranes and also serves as a precursor for the synthesis of other steroids. Cholesterol is generated in the liver and taken from the food in vertebrates. A high blood cholesterol level may lead to atherosclerosis. Indeed, both saturated and trans fats have a deleterious influence on health via influencing cholesterol levels.

3.7. PROTEINS INCLUDE A DIVERSITY OF STRUCTURES, RESULTING IN A WIDE RANGE OF FUNCTIONS

Proteins are required for nearly every dynamic activity of a living creature. Proteins' significance is highlighted by its name, which derives from the Greek word proteios, which means "first" or "first."

Proteins make up more than half of the dry mass of most cells and play a role in practically everything organisms perform. Some proteins accelerate chemical processes, while others help with defense, storage, transport, cellular communication, mobility, and structural support.

Without enzymes, the majority of which are proteins, life would be impossible. Enzymatic proteins govern metabolism by functioning as catalysts, which are chemical agents that selectively accelerate chemical processes without being consumed in the process.

Because an enzyme can repeat its action, these molecules may be regarded as workhorses that keep cells functioning by carrying out life's operations. Tens of thousands of distinct proteins exist in humans, each with its unique structure and function; proteins are the most structurally complicated molecules known.

Consistent with their different roles, proteins vary greatly in structure, with each kind having a distinct three-dimensional form.

3.7.1. Polypeptides

Proteins, as diverse as they are, are all unbranched polymers made up of the same set of 20 amino acids. Polypeptides are polymers of amino acids. A protein is a physiologically active molecule made up of one or more polypeptides that have been folded and coiled into a particular three-dimensional shape.

Amino Acid Monomers

All amino acids have the same structure. An amino acid is an organic compound that contains both an amino group and a carboxyl group.

The alpha (a) carbon atom is located in the middle of the amino acid. Its four distinct partners are an amino group, a carboxyl group, a hydrogen atom, and a variable group represented by R. The R group, also known as the side chain, varies with each amino acid.

The amino groups and carboxyl groups are all portrayed in an ionized state, as they would be at the pH observed in a cell. The side chain (R group) might be as simple as a hydrogen atom, as in the amino acid glycine, or it may be a carbon skeleton with multiple functional groups attached, as in glutamine.

The physical and chemical properties of an amino acid's side chain establish its distinctive traits, which impact its functional role in a polypeptide. One category comprises hydrophobic amino acids with nonpolar side chains.

Another category consists of hydrophilic amino acids with polar side chains. Acidic amino acids have negatively charged side chains due to the presence of a carboxyl group, which is normally dissociated (ionized) at cellular pH.

The amino groups on the side chains of basic amino acids are often positive in charge. (Note that all amino acids include carboxyl and amino groups; the phrases acidic and basic in this context solely relate to side chain groups.) Acidic and basic side chains are hydrophilic because they are charged.

Amino Acid Polymers

Since we've looked at amino acids, let's look at how they're joined together to make polymers. When two amino acids are positioned so that the carboxyl group of one is close to the amino group of the other, a dehydration process occurs, resulting in the elimination of a water molecule. A peptide bond is the resultant covalent bond.

When this process is repeated several times, a polypeptide, a polymer of numerous amino acids connected by peptide bonds, is formed. The various side chains (R groups) of the amino acids extend from this backbone.

Polypeptides can be as short as a few amino acids or as long as a thousand or more. Each polypeptide contains a distinct linear sequence of amino acids. It's important to note that one end of the polypeptide chain has a free amino group and the other has a free carboxyl group. As a result, each polypeptide has a single amino end (N-terminus) and a single carboxyl end (C-terminus). Because side chains dominate terminal groups in polypeptides of any size, the chemical character of the molecule as a whole is dictated by the kind and sequencing of the side chains. The enormous variety of polypeptides found in nature exemplifies an essential notion stated earlier: a cell may produce a wide range of polymers by joining a restricted number of monomers into varied sequences.

3.8. PROTEIN STRUCTURE AND FUNCTION

The unique actions of proteins are caused by their sophisticated three-dimensional architecture, the most basic level of which is the sequence of their amino acids. Frederick Sanger, who worked on the hormone insulin with colleagues at Cambridge University in England in the late 1940s and early 1950s, was a pioneer in establishing the amino acid sequence of proteins.

He employed agents to break polypeptides at certain points, then used chemical procedures to identify the amino acid sequence in these minute fragments. After years of work, Sanger and his colleagues were able to recreate the whole amino acid sequence of insulin. Since then, the majority of the steps required in polypeptide sequencing have been automated.

What happens once we've determined the amino acid sequence of a polypeptide, what could it teach us about the protein's three-dimensional structure (often referred to simply as "structure") and function? The terms polypeptide and protein are not interchangeable.

Even for a single polypeptide protein, the connection is similar to that of a long strand of yarn and a sweater of a certain size and form that may be created from the yarn. A functional protein is more than simply a polypeptide chain; it is one or more polypeptides that have been carefully twisted, folded, and coiled into a molecule with a distinct form.

And the amino acid sequence of each polypeptide defines the protein's three-dimensional shape under normal physiological circumstances. When a cell produces a polypeptide, the chain folds spontaneously, acquiring the functional structure of that protein.

The development of a variety of bonds between portions of the chain, which in turn depends on the sequence of amino acids, drives and reinforces folding. Many proteins have a roughly spherical form (globular proteins), whereas others have the structure of lengthy threads (fibrous proteins). There are numerous differences even within these broad groups.

The structure of a protein influences how it functions. Almost every protein's function is dependent on its capacity to detect and attach to another molecule.

An exceptionally dramatic example of the union of form and function: an antibody (a protein in the body) and the specific foreign material on a flu virus to which the antibody attaches.to and destroys. Thus, a protein's function—for example, a receptor protein's capacity to bind to a specific pain-relieving signaling molecule—is an emergent trait emerging from fine molecular orientation.

3.9. FOUR LEVELS OF PROTEIN STRUCTURE

When attempting to understand the function of a protein, studying its structure is frequently beneficial. Despite their vast differences, all proteins have three stacked layers of structures known as primary, secondary, and tertiary structures. When a protein is composed of two or more polypeptide chains, a fourth level, quaternary structure is formed.

3.9.1. Primary Structure (Linear Chain of Amino Acids)

A protein's main structure is a connected string of amino acids with a unique sequence. Consider transthyretin, a globular blood protein that carries vitamin A and one of the thyroid hormones throughout the body as an example. Transthyretin is comprised of four identical polypeptide chains, each of which has 127 amino acids.

One of these chains has been unwound to reveal its fundamental structure. One of the 20 amino acids, denoted by its three-letter acronym, occupies each of the 127 locations along the chain. The fundamental structure is similar to the arrangement of letters in a very lengthy word. If everything was up to chance, there would be 20127 possible methods to make a polypeptide chain of 127 amino acids.

Nevertheless, the particular main structure of a protein is dictated by inherited genetic information rather than random amino acid linkage. Because of the molecular composition of the backbone and the side chains (R groups) of the amino acids positioned along the chain, the fundamental structure influences secondary and tertiary structure.

3.9.2. Secondary Structure (Regions Stabilized By Hydrogen Bonds between Atoms of the Polypeptide Backbone)

Many proteins feature polypeptide chain segments that are regularly coiled or folded in patterns that contribute to the overall structure of the protein. These coils and folds, known as the secondary structure, are formed by hydrogen bonds between the repeating elements of the polypeptide backbone (not the amino acid side chains). Because the oxygen atoms inside the backbone have a partial negative charge and the hydrogen atoms connected to the nitrogen have a partial positive charge, hydrogen bonds can form between these atoms.

These hydrogen bonds are weak on their own, but because they are repeated several times throughout a relatively lengthy portion of the polypeptide chain, they can sustain a certain shape for that section of the protein. The helix, a fragile coil kept together by hydrogen bonding between every fourth amino acid, is one such secondary structure.

Although each transthyretin polypeptide only has one helix section (see tertiary structure on the following page), other globular proteins contain several helix lengths separated by nonhelical regions (see Haemoglobin on the next page).

Some fibrous proteins, such as α -keratin, the structural protein of hair, form helixes for the majority of their length. The pleated sheet is another common sort of secondary structure. As illustrated above, two or more strands of the polypeptide chain lying side by side (called strands) are joined by hydrogen bonds between sections of the two parallel polypeptide backbones in this structure.

Folded sheets form the core of many globular proteins, such as transthyretin (see tertiary structure below), and dominate other fibrous proteins, such as spider silk protein. Because of the collaboration of so many hydrogen bonds, each spider silk thread is stronger than a steel strand of the same weight.

3.9.3. Tertiary Structure (Three-Dimensional Shape Stabilized By Interactions between Side Chains)

The tertiary structure of a protein is superimposed over the patterns of secondary structure, as seen above in a ribbon model of the transthyretin polypeptide. While the secondary structure is determined by interactions between backbone elements, the tertiary structure is determined by interactions between the side chains (R groups) of the different amino acids.

One sort of contact that leads to the tertiary structure is referred described as a hydrophobic interaction, which is rather misleading. When a polypeptide folds into its functional shape, amino acids with hydrophobic (nonpolar) side chains typically end up in clusters near the protein's core, away from water.

Thus, a "hydrophobic contact" is really created by water molecules excluding nonpolar substances. When the side chains of nonpolar amino acids are close together, van der Waals's interactions aid in keeping them together.

Conversely, hydrogen bonding between polar side chains and ionic interactions between positively and negatively charged side chains aid in tertiary structure stabilization. In the aqueous cellular environment, these are all weak interactions, but their combined impact contributes to the protein's distinctive form.

Disulfide bridges, which are covalent connections, can help to strengthen the structure of a protein. When two cysteine monomers with sulfhydryl groups (SH) on their side chains are pushed close together by protein folding, disulfide bridges develop. The sulfur of one cysteine links to the sulfur of the second, and the disulfide bridge (SS) connects different parts of the protein.

3.9.4. Quaternary Structure (Association of Multiple Polypeptides, Forming a Functional Protein)

Most proteins are made up of two or more polypeptide chains linked together to form a single functional macromolecule. The overall protein structure that comes from the aggregation of these polypeptide subunits is known as the quaternary structure. The full globular transthyretin protein, seen above, is made up of four polypeptides. Another example is collagen, which is a fibrous protein with three identical helical polypeptides entwined into a bigger triple helix, providing significant strength to the long fibers. This is consistent with collagen fibers' role as connective tissue girders in skin, bone, tendons, ligaments, and other bodily components. (Collagen accounts for 40% percent of protein in the human body.) A further example of a globular protein having a quaternary structure is Hemoglobin, the oxygen-binding protein of red blood cells illustrated below. It is made up of four polypeptide subunits, two of one kind (β) and two of another type (β). Both subunits are predominantly composed of -a helical secondary structure. Each subunit has a non-polypeptide component known as heme, which contains an iron atom that binds oxygen.

3.10. SICKLE-CELL DISEASE: A CHANGE IN PRIMARY STRUCTURE

A protein's form and capacity to function can be affected by even minor changes in its fundamental structure. For example, sickle cell disease, a genetic blood disease, is caused by the replacement of one amino acid (valine) for the usual one (glutamic acid) at a specific location in the fundamental structure of Hemoglobin, the protein that transports oxygen in red blood cells.

Normal red blood cells are disk-shaped, but in sickle cell disease, the aberrant Hemoglobin molecules crystallize, causing some of the cells to distort into a sickle shape. The condition causes "sickle-cell crises," in which angular cells jam tiny blood arteries, restricting blood flow consequences.

The toll placed on such patients exemplifies how a modest alteration in protein structure may have disastrous consequences for protein function.

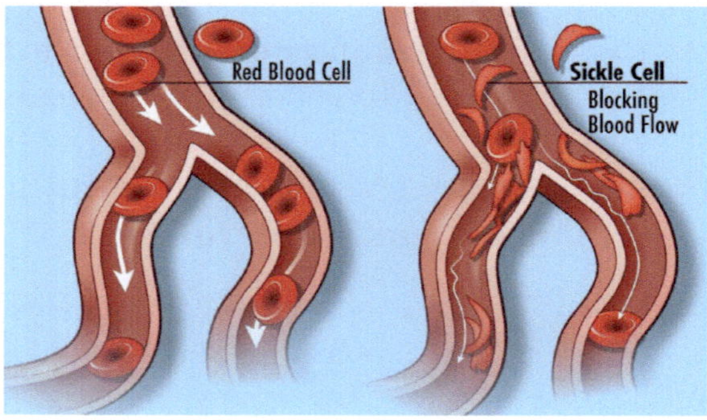

Figure 3.6. Normal blood cells (left) and the blood cells in Sickle cell disease, which does not flow through the circulatory system smoothly.

Source: Image by Wikimedia Commons

3.10.1. What Determines Protein Structure?

Each protein has a distinct form that confers a certain function. But what are the main variables that influence protein structure? You already know most of the answers: A polypeptide chain of a particular amino acid sequence can spontaneously organize itself into a three-dimensional form defined and maintained by secondary and tertiary structural interactions.

This folding generally happens while the protein is being created in the crowded environment of a cell, with the help of other proteins. Protein structure, however, is also affected by the physical and chemical circumstances of the protein's surroundings.

If the pH, salt concentration, temperature, or other features of a protein's environment change, the protein's weak chemical connections and interactions may be disrupted, causing the protein to unravel and lose its natural form, a process known as denaturation. The denatured protein is physiologically inactive due to its distorted state.

When proteins are moved from an aqueous environment to a nonpolar solvent, such as ether or chloroform, they denature; the polypeptide chain refolds such that the hydrophobic portions face outward toward the solvent.

Chemicals that break hydrogen bonds, ionic bonds, and disulfide bridges that keep a protein's structure are examples of denaturation agents. Excessive heat can also cause denaturation by agitating the polypeptide chain sufficiently to overwhelm the weak connections that maintain the structure.

Because denatured proteins are insoluble and harden during cooking, the white of an egg turns opaque. This also explains why high fevers can be deadly: Proteins in the blood can denature when exposed to extremely high body temperatures.

When a protein in a test-tube solution is denatured by heat or chemicals, it can occasionally revert to its functional structure once the denaturing agent is removed. We may conclude that the information for constructing certain shapes is inherent in the core structure of the protein.

The sequence of amino acids controls the structure of the protein—where a helix may form, where pleated sheets can exist, and where disulfide bonds can form, and so forth. But how does protein folding take place in the cell?

3.10.2. Protein Folding in the Cell

More than 10 million amino acid sequencesmm/ and three-dimensional shapes are currently known to biochemists. To determine the laws of protein folding, researchers attempted to match the basic structure of numerous proteins with their three-dimensional structure. Unfortunately, the process of protein folding is not so easy.

Most proteins are likely to travel through numerous intermediate forms before reaching a stable shape, and looking at the mature structure does not show the phases of folding necessary to attain that form. Biochemists, on the other hand, have devised ways for monitoring a protein through such phases. Chaperonins (also known as chaperone proteins) are protein molecules that aid in the proper folding of other proteins and are critical to the folding process. Chaperonins do not dictate a polypeptide's ultimate structure. Instead, they maintain the nascent polypeptide isolated from "negative stimuli" in the cytoplasm while it folds spontaneously.

The chaperonin from the bacteria *E. coli* is a massive multiprotein complex with the form of a hollow cylinder. The cavity offers a place for polypeptides to fold. Over the last decade, scientists have found molecular systems that interact with chaperonins and monitor for correct folding.

Such systems either correctly refold misfolded proteins or mark them for destruction. Polypeptide misfolding is a severe issue in cells. Misfolded proteins have been linked to a variety of disorders, including Alzheimer's, Parkinson's, and mad cow disease.

Misfolded variants of the transthyretin protein have been linked to a number of disorders, including one kind of senile dementia. Even when scientists have a perfectly folded protein in hand, establishing its exact three-dimensional structure is difficult due to the millions of atoms in a single protein molecule.

The first 3-D structures of hemoglobin and a similar protein were discovered in 1959. X-ray crystallography, which has since been employed to establish the 3-D structure of many other proteins, enabled these accomplishments.

In one current instance, Roger Kornberg and colleagues at Stanford University utilized this approach to determine the structure of RNA polymerase, an enzyme that is essential for gene expression. Nuclear magnetic resonance (NMR) spectroscopy is another approach for examining protein structure that does not need protein crystallization.

A modern method uses bioinformatics to predict the 3-D structure of polypeptides based on their amino acid sequence. Understanding protein structure and function requires a combination of X-ray crystallography, NMR spectroscopy, and bioinformatics.

3.11. STRUCTURAL FEATURES OF NUCLEIC ACIDS

Miescher isolated deoxyribonucleic acid (DNA) acid for the first time around 1870 (from white blood cells and then from salmon sperm), publishing a series of outstanding investigations on the structure of the isolated result.

Because of its presence in the cellular nucleus, he dubbed the phosphate-rich material "nuclein." When the substance's acidic nature was revealed, the term was modified to "nucleic acid."

The study of these biomolecules has revealed that they, like proteins, are biopolymers. DNA and RNA (ribonucleic acid) are chain-like polymers capable of storing and transmitting genetic information. Even though the chemistry of these molecules was carefully researched after their discovery, it has taken 75 years for the biological importance of DNA and RNA to be recognized.

Avery proposed the concept of DNA as genetic material in 1944. RNA's role in protein synthesis was discovered in 1957, despite the fact that it had previously been recognized as genetic material for certain viruses. The nucleic acids are important components, accounting for 5–15% of the dry mass of the cells. They can also be present in viruses, which are nucleic acid-protein infectious complexes that can self-replicate in the host cell.

Unlike plants and animals, viruses only have one kind of DNA or RNA, never both. Since they get their name from being separated from cellular nuclei, they can also be found in other cellular compartments (i.e. mitochondria). A nucleic acid's monomer unit is known as a nucleotide, and the polymer is known as a polynucleotide.

The monomer of DNA is a deoxyribonucleotide, while the monomer of RNA is a ribonucleotide. A nucleotide can be degraded to produce a heterocyclic nitrogenous base, pentose, and phosphoric acid.

3.11.1. Nitrogenous Bases

Purines and pyrimidines are two types of nitrogenous bases found in nucleic acids. Adenine (A) and guanine (G) are the two purines found in DNA and RNA, whereas cytosine (C) is found in both nucleic acids. Uracil (U) is found solely in RNA and replaces thymine (T), which is found in DNA.

Tautomerism creates a pH-dependent equilibrium between the keto (lactame) and enol (lactime) forms of these bases. Lactame predominates at physiological pH and is responsible for the H-bonds created between base pairs in the natural DNA molecule.

3.11.2. Nucleosides

Nucleosides are chemicals that include a nitrogenous base (purine or pyrimidine) that is covalently bound to D-ribofuranose (ribonucleosides) or 2-deoxy-Dribofuranose (deoxy-ribonucleosides) through an N—glycosidic linkage. The hemiacetal group of C-1' of pentose and the N-9 nitrogen atom of purine or N-1 of pyrimidine form these linkages. In t-RNA, a nucleoside (pseudouridine) was identified in which the C-1' of ribose is connected to the C5 of uracil.

3.11.3. Nucleotides

Nucleotides are nucleoside phosphate esters. Even though there are various kinds of nucleotides, because the phosphate might be at the 2'-, 3'-, or 5'-carbon of a ribonucleotide or the 3'- or 5'-carbon of a deoxynucleotide, naturally occurring nucleotides are often 5'-monophosphates.

In addition to monophosphate derivatives, all ribonucleosides and deoxyribonucleosides occur in cells as 5'-di- and 5'-triphosphates, which are nucleotide esters of 5'-diphosphoric and 5'-triphosphoric acid.

Thus, for adenosine, there exist three series of 5'-phosphorylated nucleosides: adenosine monophosphate (AMP), adenosine-diphosphate (ADP), and adenosine triphosphate (ATP). These compounds' phosphate groups are indicated by α, β, γ.

The nucleosides 5'-diphosphoric and 5'-triphosphoric acids are relatively strong acids that liberate three and four protons from the phosphate groups, respectively. These acids' phosphate groups form compounds with the bivalent ions Mg^{2+} and Ca^{2+}.

Because of the comparatively high Mg^{2+} content in the cytoplasm, nucleoside 5'-di- and 5'-triphosphates occur as magnesium complexes in healthy cells. The hydrolysis of the phosphate group of the triphosphate derivative is a significant source of chemical energy that biological systems use to function.

3.12. THE COMPONENTS OF NUCLEIC ACIDS

Nucleic acids are macromolecules that occur in the form of polymers known as polynucleotides. Each polynucleotide, as the name implies, is made up of monomers known as nucleotides. In general, a nucleotide is made up of three components: a nitrogen-containing (nitrogenous) base, a five-carbon sugar (a pentose), and one or more phosphate groups.

Each monomer in a polynucleotide contains only one phosphate group. A nucleoside is the component of a nucleotide that does not include any phosphate groups.

Consider the nitrogenous bases first when constructing a nucleotide. Each nitrogenous base has one or two rings containing nitrogen atoms. (They are referred to as nitrogenous bases because nitrogen atoms prefer to soak up H+ from solution, serving as bases.) Nitrogenous bases are classified into two groups:

Purines and pyrimidines: A pyrimidine is a carbon and nitrogen atom-containing six-membered ring. Cytosine (C), thymine (T), and uracil are members of the pyrimidine family (U). Purines have a six-membered ring fused to a five-membered ring and are bigger.

Adenine (A) and guanine (G) are the purines (G). The chemical groups connected to the rings of different pyrimidines and purines vary. Adenine, guanine, and cytosine can be found in both DNA and RNA; however, thymine can only be found in DNA and uracil in RNA. Let's now add some sugar to the nitrogenous base.

The sugar in DNA is deoxyribose, while the sugar in RNA is ribose. The main distinction between these two sugars is that deoxyribose does not have an oxygen atom on the second carbon in the ring. Deoxyribose is derived from this.

The sugar carbon numbers of a nucleoside or nucleotide have a prime (') after them to separate them from the numbers used for the ring atoms of the connected nitrogenous base. Thus, the second carbon in the sugar ring is referred to as the 2' ("2 prime") carbon, while the carbon that protrudes from the ring is referred to as the 5' carbon. So far, we have constructed a nucleoside (nitrogenous base plus sugar).

To finish the nucleotide, we connect a phosphate group to the 5' carbon of the sugar. The molecule is currently known as a nucleoside monophosphate or nucleotide.

3.12.1. Nucleotide Polymers

One can now see how these nucleotides are joined to form a polynucleotide. A phosphodiester linkage connects adjacent nucleotides via a phosphate group that connects the sugars of two nucleotides. The outcome of this bonding is a backbone with a repeating pattern of sugar-phosphate units. (It should be noted that the nitrogenous bases do not form part of the backbone.) The polymer's two free ends are noticeably different from one another. The 5' end has a phosphate connected to a 5' carbon, while the 3' end has a hydroxyl group attached to a 3' carbon; these are referred to as the 5' end and the 3' end, respectively.

A polynucleotide contains built-in directionality along its sugar-phosphate backbone, ranging from 5' to 3'. Similar to a one-way street Appendages made up of nitrogenous bases run the length of this sugar-phosphate backbone. The sequence of nucleotides along with a DNA (or mRNA) polymer is unique to each gene and supplies the organism with extremely particular information.

Because genes can be hundreds or thousands of nucleotides long, the number of potential base sequences is practically infinite. The significance of a gene to the cell is encoded in its particular sequence of four DNA bases.

The sequence 5'-AGGTAACTT-3', for example, implies one thing, but the sequence 5'-CGCTTTAAC-3' means something other. (Of course, whole genes are far longer.) The linear arrangement of nucleotides in a gene determines the amino acid sequence—the main structure of a protein, which determines its three-dimensional shape and function inside the cell.

3.12.2. The Structures of DNA and RNA Molecules

RNA molecules are often found as single polynucleotide chains. DNA molecules, on the other hand, have two polynucleotides, or "strands," that spiral around an imaginary axis to create a double helix.

The two sugar-phosphate backbones run in opposing 5' S 3' directions; this arrangement is known as antiparallel and is similar to a divided highway. The sugar-phosphate backbones lie on the outside of the helix, while the nitrogenous bases are paired within.

Hydrogen bonding between the linked bases holds the two strands together. Most DNA molecules are hundreds or even millions of base pairs long. One lengthy DNA double helix has many genes, each of which represents a different portion of the molecule. Only a subset of the bases in the double helix is compatible with one another. Adenine (A) is always associated with thymine (T), while guanine (G) is always associated with cytosine (C). We would know the sequence of bases along the other strand of the double helix if we read the sequence of bases along one strand. If a length of one strand contains the base sequence 5'-AGGTCCG-3', then the identical stretch of the other strand must have the sequence 3'-TCCAGGC-5', according to the base-pairing laws.

The two strands of the double helix are complementary, with one being the expected opposite of the other. This property of DNA allows for the formation of two identical copies of each DNA molecule in a dividing cell.

When the cell divides, copies are transmitted to the daughter cells, resulting in genetically identical daughter cells. Thus, the structure of DNA explains how it transmits genetic information whenever a cell reproduces.

Complementary base pairing can also occur between sections of two

RNA molecules or between two nucleotide lengths in the same RNA molecule. Indeed, base pairing inside an RNA molecule permits it to assume the specific three-dimensional structure required for its activity.

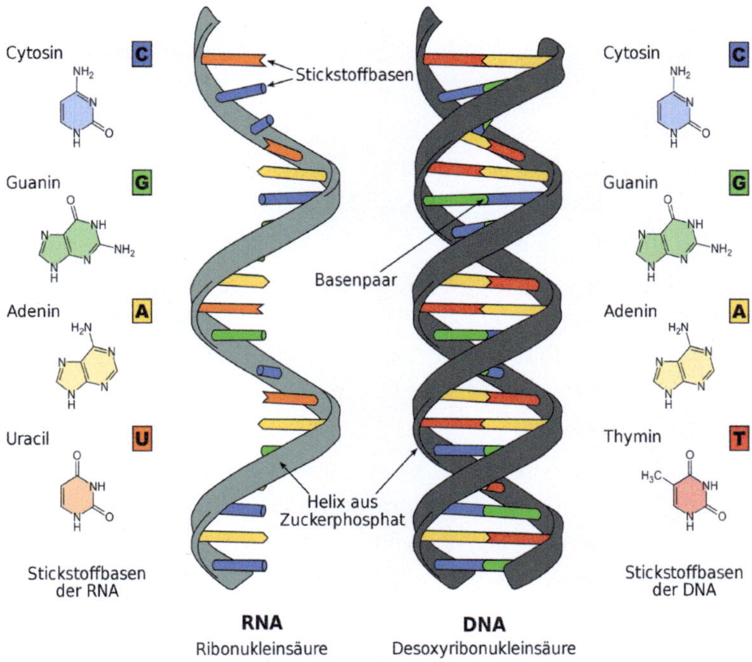

Figure 3.7. Comparison of a single-stranded RNA and a double-stranded DNA.

Source: Image by Wikimedia Commons

Consider the form of RNA known as transfer RNA (tRNA), which transports amino acids to the ribosome during polypeptide synthesis. A tRNA molecule is around 80 nucleotides long. Its functional structure is the consequence of base pairing between nucleotides, with complementary regions of the molecule running antiparallel to one another. Adenine (A) pairs with uracil (U) in RNA; thymine (T) is not present in RNA. Another distinction between RNA and DNA.DNA is that DNA nearly always exists as a double helix, whereas RNA molecules have a more varied form. This variation emerges because the degree and position of complementary base pairing inside an RNA molecule change across various kinds of RNAs in the double helix.

3.12.3. DNA and Proteins as Tape Measures of Evolution

Humans are accustomed to considering similar features, such as hair and milk production in animals, to be proof of common ancestors. Because we now know that DNA contains heritable information in the form of genes, we can see that genes and their products (proteins) reflect an organism's genetic history.

The nucleotide sequences in DNA molecules are handed down from parents to children, and these sequences define the amino acid sequences of proteins. Siblings show more DNA and protein similarities than unrelated individuals of the same species.

If the evolutionary theory of life is correct, we should be able to apply the idea of "molecular genealogy" to species relationships: One should anticipate two species that appear to be closely related based on fossil and anatomical evidence to share more DNA and protein sequences than more distantly related species.

That is, in fact, the situation. A comparison of the polypeptide chain of human hemoglobin with the comparable hemoglobin polypeptide in other vertebrates is one example. Humans and gorillas vary in only one amino acid in this 146-amino acid chain, but humans and frogs differ in 67 amino acids. A new tape measure has been introduced to the toolset used by scientists to estimate evolutionary kinship.

3.13. CONCLUSION

This chapter discussed the different macromolecule structure and their functions. It also discussed the synthesis and breakdown of the polymers. This chapter provides highlights on the diversity of the polymers. Towards the end of the chapter, the structure of the proteins and their functions have been discussed. It also discussed the four levels of protein structure, along with the components of the nucleic acids.

REFERENCES

1. Bawn, C., 1978. Macromolecules structure and properties vol 2. *Endeavour*, [online] 2(1), p.48. Available at: <https://www.sciencedirect.com/science/article/abs/pii/016093277890042X?via%3Dihub> [Accessed 29 June 2022].
2. Ledbetter, J. and Seaman, W., 1982. The Lyt-2, Lyt-3 Macromolecules: Structural and Functional Studies. *Immunological Reviews*, [online] 68(1), pp.197-218. Available at: <https://onlinelibrary.wiley.com/doi/10.1111/j.1600-065X.1982.tb01065.x> [Accessed 29 June 2022].
3. Rehder, D., Nelson, R. and Borges, C., 2009. Glycosylation status of vitamin D binding protein in cancer patients. *Protein Science*, [online] 18(10), pp.2036-2042. Available at: <https://onlinelibrary.wiley.com/doi/10.1002/pro.2144> [Accessed 29 June 2022].
4. Resources.finalsite.net. n.d. *The Structure and Function of Large Biological Molecules*. [online] Available at: <https://resources.finalsite.net/images/v1560347370/ucfsdorg/ymnhuceddsx8qems1u5q/Chapter5-TheStructureandFunctionofLargeBiomolecules.pdf> [Accessed 29 June 2022].
5. Srivastav, R., Sharma, R., Tandon, S. and Tandon, C., 2019. Role of DHH superfamily proteins in nucleic acids metabolism and stress tolerance in prokaryotes and eukaryotes. *International Journal of Biological Macromolecules*, [online] 127, pp.66-75. Available at: <https://www.sciencedirect.com/science/article/abs/pii/S0141813018358434> [Accessed 29 June 2022].
6. W. Thrich, K., 2007. Biological Macromolecules: Structure Determination in Solution. *Encyclopedia of Magnetic Resonance*, [online] Available at: <https://doi.org/10.1002/9780470034590.emrstm0028> [Accessed 29 June 2022].

CHAPTER 4

Synthetic Versatility and Structural Modularity in Organometallic Polymers

CONTENTS

The chapter begins with an introduction to synthetic versatility and its relationship with polymerization. With discussing polymerization, the same process incorporated for Organometallic Monomers and later Olefin and Alkyne is discussed.

Other methods employed in obtaining Organometallic Polymers are also discussed such as the Substitution and Condensation Reactions and a slightly distinct method like Electro-Polymerization. Another reaction called the Alkyne Cross-Coupling Reaction is discussed which is an example of homo-polymerization of AA-type monomers.

Ring-Opening Polymerization is the last type of polymerization discussed under this heading, after it the methods of Copolymerization of Organometallic with Organic Monomers are listed with details of the same. Under this, Alkene Polymerizations and are elaborated thereon. The ending of the chapter deals with the subsequent research and discussion pertaining to various processes of organometallic polymer synthesis.

Development towards using different methods such as the Bis(Carbene)-Based Organometallic Polymers, its failure and success are certain topics this chapter touches upon.

4.1. INTRODUCTION

There are two key areas of general discussion when it comes to tunability within a functional material: (1) versatile synthetic techniques and (2) the breadth of suitable structural features within the monomeric scaffold. These two concerns rarely avoid overlap to a certain extent, yet a universal solution to both within any polymer design is non-trivial.

Further, synthetic versatility can be broken down into having either multiple access routes in order to obtain the general monomer template or having a versatile and multifunctional monomer that can partake in more than one type—or in mechanistically distinct—polymerizations (e.g., co-polymerizations) with a high level of control.

The polymerization technique, the stability of the metal core, and the position of the metal's center are all essentially dependent on structural modularity (i.e., whether main- or side-chain metal incorporation). However, assuming that the reaction conditions are generally compatible with the functional groups needed, the monomer design should allow for their installation. There are a couple of methods through which main-chain organometallic polymers can be prepared synthetically. This chapter provides an introduction to the methods that involve homo- and copolymerization of organometallic monomers, copolymerization of organometallic with organic monomers, and copolymerization of inorganic reagents with organic monomers in such a way that the bonds which are formed to the metal involved are the ones that lead to polymer formation.

Years of research have gone into each method of polymer synthesis. Although many nuanced details may be discussed in greater depth, for the purposes of this chapter, just a high-level overview will be provided.

4.2. POLYMERIZATIONS OF ORGANOMETALLIC MONOMERS

Functionalized organometallic compounds are frequently used as a starting point for structurally simple metal-containing polymers. It's no surprise that the huge body of information available for the small-molecule synthesis of metal complexes is frequently used and encompasses numerous subclasses of macromolecules.

Many groups have succeeded in developing ligands with reactive sites either distal or proximal to the metal's point of contact, such as arenes containing halogens poised for substitution. The binding affinity between the ligand and the transition metal contained in the polymer chain is frequently used to determine the stability of the organometallic polymer.

4.2.1. Olefin and Alkyne Polymerization

The polymerization of vinyl ferrocene (1) by Arimoto and Haven (Figure 4.1) is considered as the point of inception of organometallic polymers. Since that report, the surge of additional techniques and monomer structures befitting for alkene polymerizations has strengthened considerably.

Olefin polymerization approaches are used to obtain side-chain organometallic polymers, generally even though many examples of main-chain systems have also been achieved. This method has the advantage that, if the organometallic moieties are stable, nearly any robust alkene or alkyne metathesis reaction that is compatible with organic monomers is also compatible with organometallic versions.

While alkyne metathesis has received less attention, one of its significant properties is the capacity to produce an organometallic polymer with a fully conjugated all-carbon backbone.

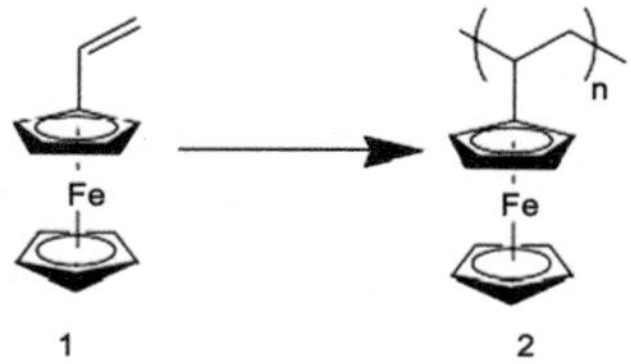

Figure 4.1. The polymerization of vinyl ferrocene.

Source: Image by Springer

4.2.2. Substitution and Condensation Reactions

The excellent methods for generating organometallic polymers are the S_NAr approach and polycondensation reactions. Keeping in mind the structural complexity and control, perhaps the most exemplary organometallic polymers are alternating bimetallic polymers. As given in Figure 4.2, isolation of Fe-complex (3) and subsequent reaction with cationic Ru complex (4) gave an alternating bimetallic polymer with excellent control. There are a number of advantages to taking this approach.

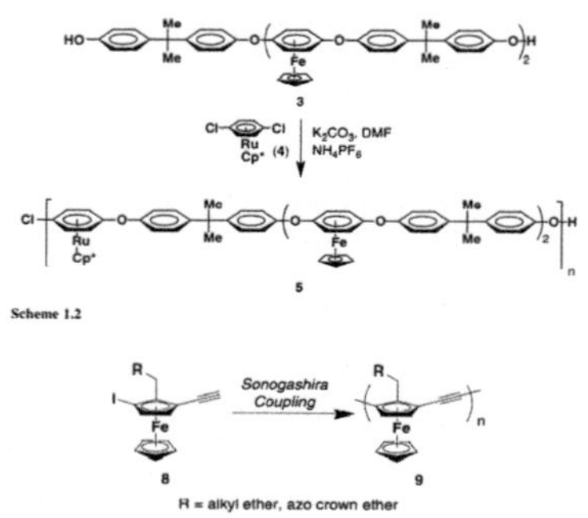

Figure 4.2. Isolation of Fe-complex (3) and subsequent reaction with cationic Ru complex (4) gave an alternating bimetallic polymer.

Source: Image by Springer

To begin, each organometallic monomer can be produced and described separately. Second, by using a hetero-coupling copolymerization procedure, full control over the alternating location of each metal-containing moiety within the polymer chain is achieved. Finally, because metallocene chemistry has a large number of metal combinations, the system's modularity should be high.

4.2.3. Electro-Polymerization

Another appealing method for generating metal-containing polymers from discrete organometallic monomers is electrochemical polymerization. Most of the time, the polymers formed are side-chain organometallics rather than main-chain organometallics (Figure 4.3,6).

However, there are a few examples of main-chain organometallic polymers made by electropolymerization. To make polymers like those seen in Figure 4.3,7), Constable used a functionalized Ru(terpy)2 complex with electro-polymerizable thiophenes on the periphery.

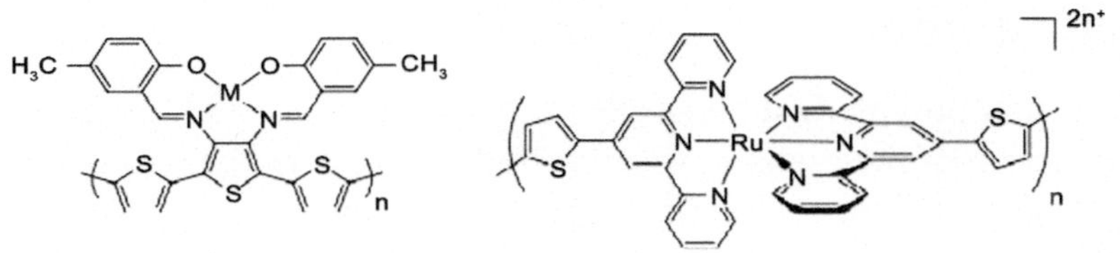

6 7

Figure 4.3. Examples of polymers obtained via electro-polymerization of organometallic monomers.

Source: Image by springer

4.2.4. Alkyne Cross-Coupling Reactions

So far, all of the photopolymerizations discussed have been with AA-type monomers. Alkyne cross-coupling techniques provide a useful way to manage AB-type monomer polymerizations. Plenio, for example, created highly functionalized monomers (8) with an aza crown ether, as well as an iodo and ethynyl group ready for homopolymerization (Figure 4.4).

It is important to note that Plenio and coworkers earlier published polymers with remarkable optical activity and architectures that were also based on 9 (Figure 4.4). The value of modularity is demonstrated by the synthesis of polymers with widely different potential applications that all come from the same synthetic design.

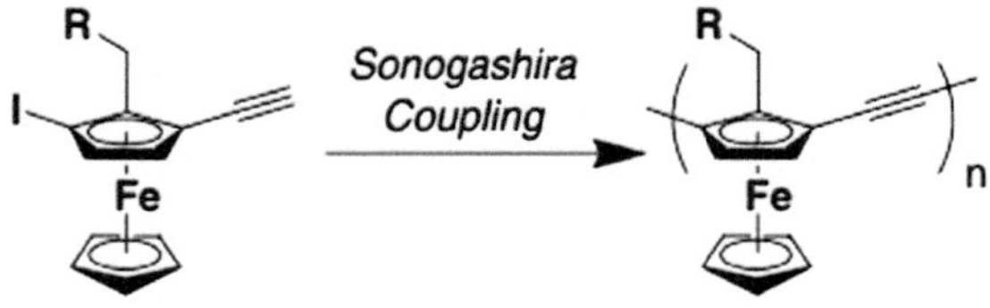

8 9

R = alkyl ether, azo crown ether

Figure 4.4. Alkyne cross-coupling reactions.

Source : Image by springer

4.2.5. Ring-Opening Polymerization

Ring-opening polymerization (ROP) offers a wide range of applications in the production of main-chain organometallic polymers. The transformation from 10 to 11 (Figure 4.5), first reported by Rauchfuss and extensively developed by Manners and coworkers, has been optimized to include a variety of polymerization conditions, including thermal, anionic, photo, and metal-mediated polymerizations; both solution and solid-state polymerization have also been reported.

The capacity to produce monomers of variable functionality has aided in overcoming any inherent solubility limits. Molecular weights on the order of 106 Da have been obtained. Hydrocarbon, sulfur, boron, tin, germanium, phosphorus, and silicon bridges, as well as other segments generated from block co-polymerizations, have all been used as bridging groups.

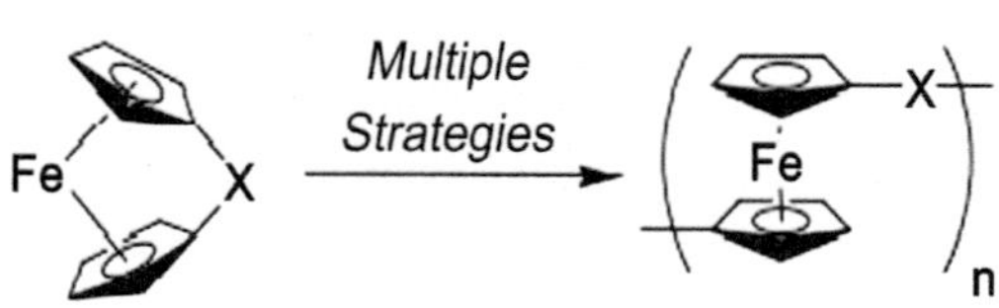

10 11

Figure 4.5. Ring-opening polymerization.

Source: Image by springer

4.3. COPOLYMERIZATION OF ORGANOMETALLIC WITH ORGANIC MONOMERS

4.3.1. Alkene Polymerizations

Organometallic and organic substrates can be copolymerized using the majority of alkene polymerization reactions. Since Pittman first described the radical polymerization of acrylate functionalized metal-arene complexes in the early 1970s, a lot has happened in this field. Although some monomers homopolymerized without difficulty, most macromolecular products with significant molecular weight required the addition of an organic comonomer. Organometallic substrates have also been used to make mixed-metal copolymers, in addition to organic comonomers. This method is also compatible with modern polymerization techniques that are highly controlled. The living free radical polymerization of vinyl ferrocene with styrene, for example, was reported by Frey and colleagues.

4.3.2. Substitution and Condensation Reactions

Alike the all-organometallic polymers that have been discussed previously, substitution and condensation reactions are also widely used for the copolymerization of organic and organometallic monomers. The S_NAr approach is employed almost exclusively in the application of the metallocene as an electrophile since the halogen is activated by metal complexation to the arene. Because the report by Segal on the polymerization of CpRu complexed with p-dichlorobenzene (12) along with various bis(phenate) ions (Figure 4.6), this approach has been developed to incorporate a wide range of structures. In the organic comonomers, the greatest structural variations can be found however, polyhalogenation of the metallocene provides a pathway for structural tuning as well.

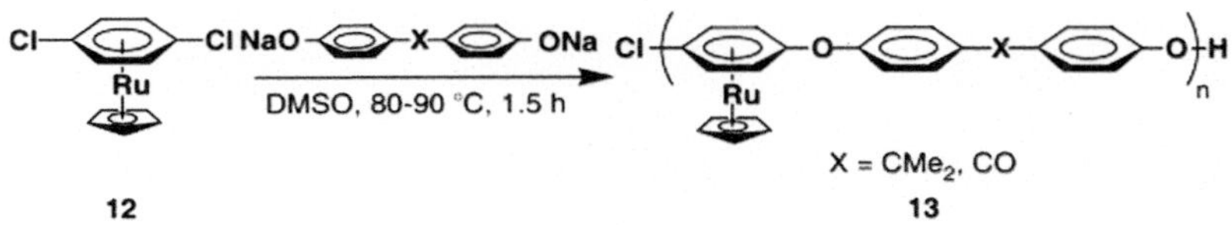

Figure 4.6. Image showing substitution and condensation reaction.

Source: Image by springer

Dembek and coworkers have demonstrated the ability to generate highly branched materials using tri- and tetra-chlorobenzene complexes, extending their work with Cp Ru-dichlorobenzene complexes. This last example demonstrates the molecular and architectural complexity that can be achieved with this route using only simple synthetic manipulations like polyhalogenation of arenes and metal complexation.

There are numerous examples of metal-arene complexes with nucleophiles poised for polycondensation to produce polyesters, amides, and imines, among other things. Polycondensation reactions involving metallocene diacid chloride variations have been studied since 1961 with various linkers such as 1,4-hydroquinone and p-phenylenediamine.

The organometallic monomer can be used as either a nucleophilic or an electrophilic partner in polycondensation reactions. Figure 4.7 shows some examples of each. Jin and Kim demonstrated that combining the phenylenediamine-Cr(CO)3 complex (14) with tere-phthaloyl chloride results in a good yield of the corresponding copolymer (15). Wright and Lowe-Ma used a Cr complex of terephthalaldehyde (16) and m-phenylenediamine to perform dehydration reactions to form polyimines, which is similar to this example. Although the solubility of this particular polymer was low, the method could easily be modified to include monomers with higher solubilizing abilities.

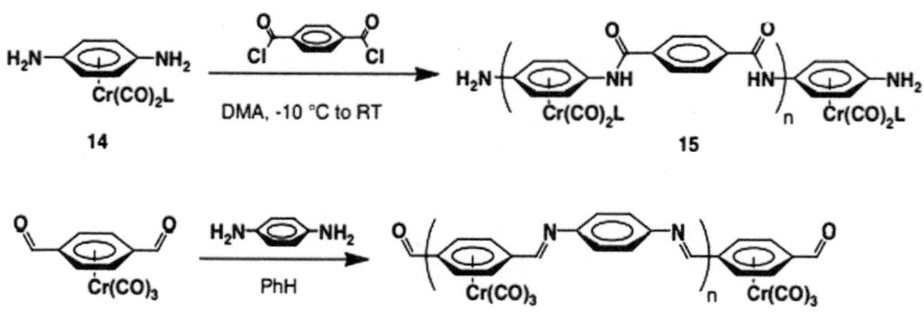

Figure 4.7. Reaction of phenylenediamine-$Cr(CO)_3$ complex with tere-phthaloyl chloride to give the corresponding copolymer.

Source: Image by springer

4.3.3. Cross-Coupling Reactions

Others have achieved an all-carbon backbone using standard cross-coupling techniques to install conjugated linkers. The metal-arene complex has been used in both roles (individually and dually) of the cross-coupling in these cases.

Under standard conditions, Wright used the electrophile metallocene (18) in combination with Stille reagent 19 to produce the corresponding highly conjugated organometallic polymer (Figure 4.8). Despite their low molecular weight, these polymers had poor solubility, as one might expect given their linear rigid framework (ca 7.8 kDa).

Alternatively, Chujo and colleagues used the 1,4-diethynylbenzene chromium complex (21) in combination with dibromothiophenes (22) to make a functionalized thienylene-containing copolymer (23) under Sonogashira conditions. Polymers with molecular weights ranging from 13.5 to 24.4 kDa (PDIs = 3.2–3.6) were obtained in these studies, and comparative UV-Vis spectroscopy confirmed extensive p-delocalization.

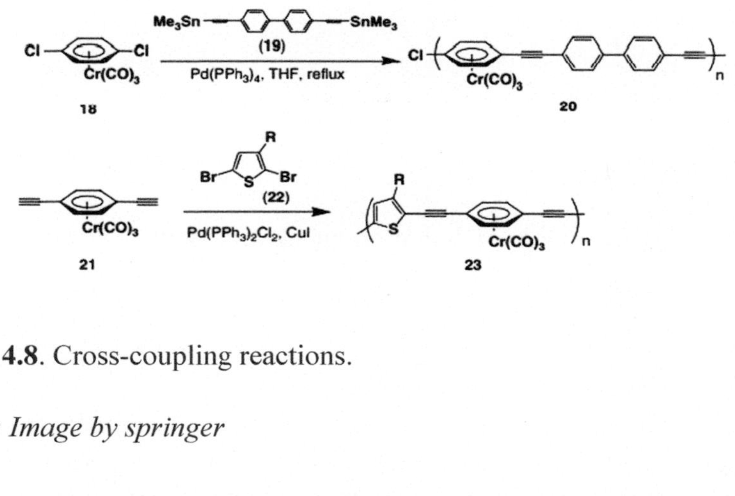

Figure 4.8. Cross-coupling reactions.

Source: Image by springer

4.4. POLYMERIZATIONS INVOLVING METAL-BINDING EVENTS DURING POLYMERIZATION

4.4.1. Metal-Containing Polyenes

Metal-containing polyenes are a fascinating type of main-chain organometallic polymer that has been studied since the 1970s. These systems are known for their rigid-rod structures and electronic communication over a highly delocalized -system, which results in interesting optical properties.

Much optimization and development have occurred since Hagihara's initial report on Pd- and Pt-containing polymers. The use of a metal(II) halide (24) in combination with an α,ω-diyne is a common synthetic route (25). As shown in Figure 4.9, a variety of linkers have been used, as well as many transition metals and ligand effects at the metal center.

Simple functionalization of the arene linkers allows for electronic and solubility tuning. Metal-alkyne linkage in the polymerization step has also been used to produce mixed-metal bimetallic.

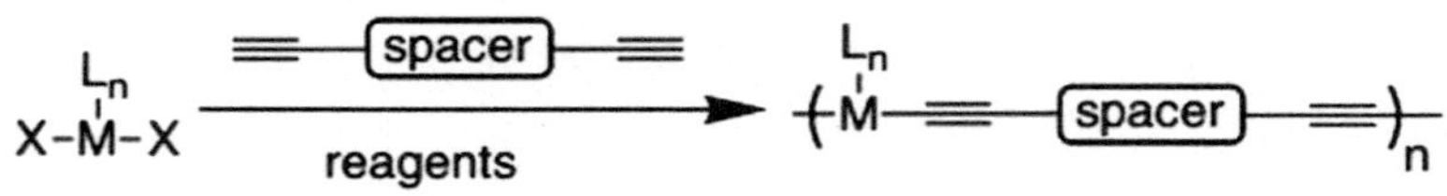

24

25

Figure 4.9. Metal-containing polyenes.

Source: Image by springer

4.4.2. Coordination Polymers

Given the vast body of knowledge about neutral donor ligands in metal complexes that is almost taken for granted, it's no surprise that many researchers have used these moieties in the design of organometallic polymers.

The task at hand appears to be as simple as attaching two known ligands to either end of a long and/or rigid linker. While this is essentially the overall plan, creating macromolecular materials in this manner is simple in theory but difficult in practice. Phosphines, mono-, bidentate-, or tridentate amines, ethers, imines, nitriles, and thio compounds have all been used as donor moieties in the past.

High binding affinities are required to produce high molecular-weight materials, and the inherent tendency toward depolymerization makes it difficult to characterize the polymers, especially in dilute solutions used in GPC, UV-Vis, and mass spectroscopy. According to Sijbesma's reports, high-molecular-weight materials can be made using difunctional bis(phosphines). The researchers discovered that simply mixing Pd(II) or Pt(II) salts with a bis(phosphine) produced macromolecular materials in these experiments. Despite the fact that a common concern with labile ligand coordination polymers is their inherent lack of structural integrity and strength, Sijbesma's polymers were sufficiently stable to form fibers.

Examples reported by Puddephatt use Au(I) capped bis(acetylides) in combination with bis(phosphine)s to produce macromolecular coordination polymers. Surprisingly, they demonstrated the coordination of AuCl to the bis(phosphine), which was followed by a reaction of the bis(acetylide).

Coordinating amines are also widely used. Before the report submitted by Puddephatt, Takahashi had used pyridines which were linked through hydrocarbon chains in order to coordinate between metal centers providing cationic metallo- polyenes of interesting properties as well as structures. However, when the same process is used by employing amine, it is more common that each binding pocket is made up of a di- or triamine.

Difunctional linkers that are derived from terpyridyl (terpy) ligands, for example, provide an extremely high binding affinity and many derivatives that are structural in nature and are further known to be utilized for polymer formation.

In such cases, a range of conjugated and otherwise functionalized spacers have been used to connect two terpy moieties. Rowan and Weder created metal binding sites on the ends of p-phenylene-ethynylene oligomers using a pyridine-based chromophore functionalized with two benzimidazoles. Supramolecular polymers were created in the presence of Zn(II) or Fe(II).

Diamine linkers, which have also been used to make organometallic coordination polymers, are closely related. Rehahn's work exemplifies some of the key characteristics of this subclass of polymers. Polymers containing Cu(I) or Ag(I) were synthesized using phenanthrolines connected by rigid conjugated spacers. Because the macromolecular structure of these polymers can be controlled by the solvent, their properties are quite intriguing.

A linear structure with a "classical" polymeric form is obtained in noncoordinating solvents, and concentration dependence of molecular weight and structure is not observed. Aggregates are thought to form in more coordinating solvents, which appeared as cyclic oligomers at high dilution.

4.5. RESEARCH AND DISCUSSION

4.5.1. New Approach to Modular Difunctional Monomers

The various methods for producing organometallic polymers that have been described thus far have several key characteristics. Various polymerization protocols have been presented, as well as multiple transition metals that can be incorporated.

Coordination polymers have a dynamic behavior that can be optimized for controlled reversible polymerizations, despite the fact that they are often too labile for practical implementation in devices. Metal-arene and metallocene polymers can be main-chain or side-chain organometallic macromolecules, with hybrids of the two being investigated on occasion.

Metal moieties can be incorporated before, after, or during polymerization (or copolymerization), with equal flexibility in the polymerization mode. Based on a review of the literature on organometallic polymers, it appears that a universal monomer scaffold with high tunability would provide the following benefits:

1. Monomer synthesis should be simple and straight-forward with high overall yield, and the general structure should have multiple access routes to facilitate modification;
2. The steric environment around the metal should be simple to manipulate, predictable, and modular;
3. A wide range of transition metals should be compatible without changing the polymerization method significantly;
4. Functionalization sites should be visible and easily manipulated to control both physical and electronic features, preferably independently of one another;
5. Polymers should have "bench stability" toward moisture and air while exhibiting dual controllable dynamic behavior at he metal center;
6. The polymerization protocol should be robust, proceed under mild conditions, and not require the use of inert atmosphere or dry solvents; and
7. High molecular weight polymers should be used.

   Many of the most efficient organometallic polymer synthesis strategies (in terms of controllability and molecular weight) rely on highly developed polymerization reactions that were originally designed for all-organic substrates.

Many of the polymers created under these conditions are side-chain polymers with limited communication between metal centers. Due to the sensitivity of either the organometallic moiety or the reactivity of the comonomers, polymerization protocols frequently necessitate the use of an inert atmosphere and dry solvents.

Because the point of attachment and synthetic focus in metallocene-based systems is primarily on the functionalization of the arene, the same sites used for electronic and steric tuning are frequently coincident with those used for attachment of polymerizable functional groups.

Although for a S_NAr approach there are various ways of structural tuning, one exception is that the organometallic complex ought to almost always be designated as the electrophilic partner due to the substitution reaction relying on activation of the arene-halide bond after metal complexation.

This last concern is resolved for polycondensation reactions and when precise stoichiometric control and pristinely pure reagents are used, excellent control over molecular weight can be achieved.

Despite the daunting list of criteria for a highly modular design, it appeared that an organometallic scaffold could potentially lend itself to a solution to such extreme demands.

We imagined using heterocyclic carbenes in a novel way to create organometallic polymers with all of the characteristics listed above. Bis(carbene)s were rarely used in the synthesis of macromolecular organometallic materials prior to our contributions, with the exception of labile Ag-based aggregates.

This was surprising given heterocyclic carbenes' desirable properties for polymer formation, such as structural diversity and tunable affinities toward nearly every transition metal. Our approach would necessitate the formation of a monomer with two facially opposed heterocyclic carbene moieties linked by a rigid organic framework, which would allow us to control the metalation site and keep intramolecular chelation at bay.

Years of effort resulted in N-heterocyclic carbenes that are stable and isolable. The task of preparing a bis(carbene) target, on the other hand, seemed more difficult, especially given the history of difficulties in preparing such a class of difunctional ligands. Current research focuses on the design and synthesis of novel structures, as well as the incorporation of stable bis(carbene) moieties into macromolecules that are designed to perform useful tasks.

4.5.2. Difunctional Heterocyclic Carbenes as Linkers

The original series of bis(carbene) structures focused on a practical synthesis that can be used to demonstrate modularity primarily of the conjugated linker between the metal centers. Tetraamino arenes with various aromatic frameworks, such as benzo, biphenyl, and dioxin-based chromophores, were used to achieve this (Figure 4.10).

The tetraamines (30) were cyclized with formic acid to produce each bis(imidazole) (31) in high yield and purity. The bis(imidazoles) were first treated with NaH in refluxing PhCH3 before the electrophile was introduced, resulting in fourfold alkylation.

At this point in the reaction, adding dimethylformamide (DMF) as a cosolvent aided the dissolution of partially alkylated intermediates, and the final products (32) were precipitated cleanly from the cooled reaction mixture in good to excellent yields. The installation of N-substituents was limited to primary halides in this protocol, which was both quick and high yielding. Finding additional pathways to the monomer template was necessary to increase the number of possible N-substituents.

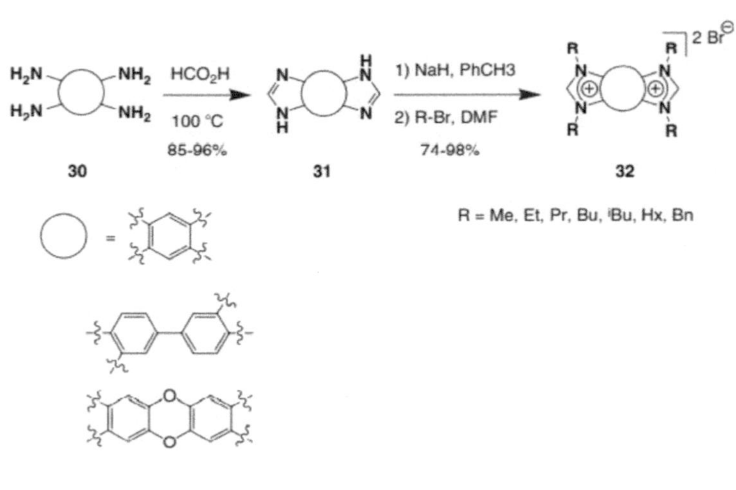

Figure 4.10. Difunctional heterocyclic carbenes as linkers.

Source: Image by springer

A two-step high-yielding protocol (Figure 4.11) involving a fourfold aryl amination to give tetraamines (34), using a modified version of Harlan's procedure, was used to install larger substituents. Our initial focus was on bulky aliphatic amines, but we later expanded our efforts to include functionalized arylamines as well. Although oxidative instability is common in tetra-amino arenes, we discovered that large alkyl groups (e.g., t Bu, t Oct, Ad) significantly reduced the rate of oxidation. Under normal conditions, these compounds can be stored for days. More reactive tetraamines, on the other hand, were isolated as hydrochloride salts and found to be highly resistant to oxidation. To provide the bis(azolium) salts, triethyl-orthoformate and HCl were used to successfully close the ring. (35). In the cases where increased solubility is desirable, simply performing the double ring closure in the presence of HBF_4 in place of HCl provided the more soluble tetrafluoroborate salts. In cases involving very large N-substituents, deprotonation gives the stable bis(carbene)s (36) which can be isolated and stored indefinitely. Previously, as it has been mentioned desymmetrization of monomers is often a focal point for accomplishing structural complexity. In addition to symmetric bis(azolium) salts, we discovered that a short sequence could be used to obtain high yields of asymmetric monomers. Finally, with complete regiocontrol, structural diversity was achieved by using bis(carbene) scaffolds with two or three different R-groups and varying heteroatoms.

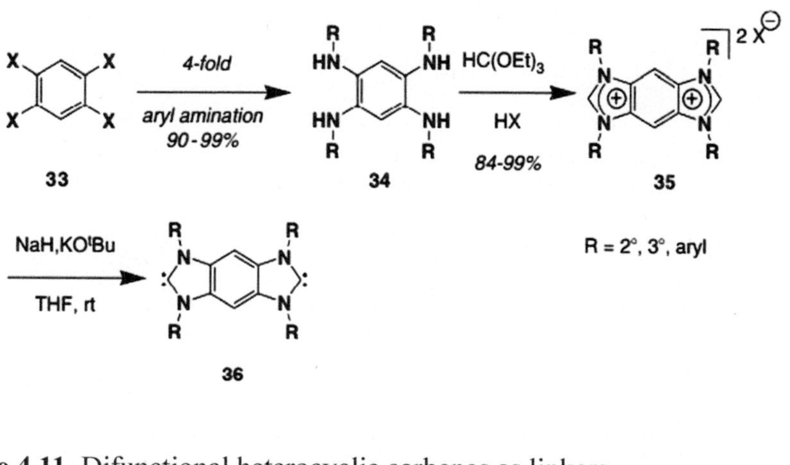

Figure 4.11. Difunctional heterocyclic carbenes as linkers.

Source: Image by springer

This was achieved by using dichloro-dinitrobenzene (37) and virtually any primary amine of choice in a double-SNAr reaction, or two sequential SNAr substitutions. This effectively regio-specifically installs the first two R-groups. The N,N′-disubstituted benzimidazoles are obtained in excellent overall yield and purity using an in-situ reduction cyclization protocol optimized in our laboratories. Bis(azolium) salts with varying substitution patterns are produced by high-yield alkylation.The diamine (38) is produced by substituting two equivalents of an amine in refluxing ethanol as it precipitates from the reaction mixture in the first route shown in Figure 4.12. When installing R-groups that aren't compatible with alkylation via nucleophilic attack on alkyl halides, this method is the best option. Alkylation can be used to produce the bis(azolium) salt with two different R-groups regio-specifically after reduction and cyclization to give 39. We discovered that mono-alkylation of 39 could be controlled to produce high yields of monoazolium salt in subsequent research. (41).

As will be discussed below, this compound proved to be a useful asymmetric building block in the preparation of alternating bimetallic polymers. After a second alkylation of 41, the bis(azolium) salt (42) was formed, which had three different R-groups. In cases where the desymmetrizing substituents are incompatible with alkylation, three different R-groups can be installed. This is accomplished by controlling the temperature of the SNAr reaction.

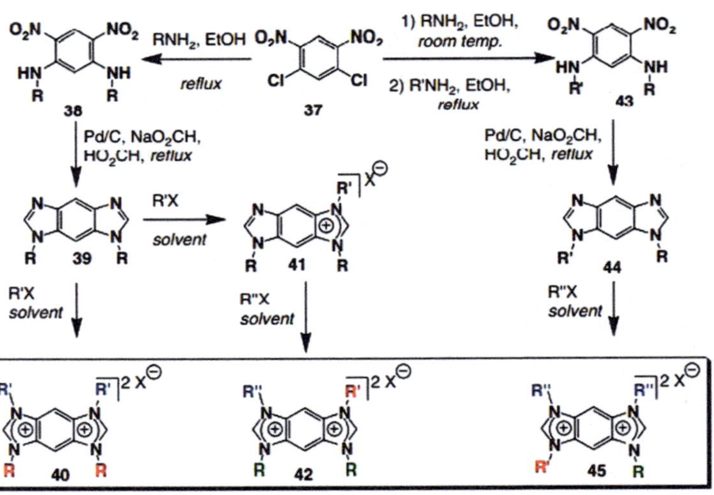

Figure 4.12. Substitution of two equivalents of an amine in refluxing ethanol provides the diamine.

The monoamine product was nearly quantitatively yielded after 37 was treated with an amine in EtOH at room temperature. The second substitution was carried out with high precision and yield, effectively desymmetrizing the monomer template (43). Reduction and cyclization of 44, followed by alkylation, yielded bis(azolium) (45), which is complementary to 42 in the R-group structure.

Source: Image by springer

In SNAr reactions, heteroatom variation was achieved by using NaSH as a nucleophile (Figure 4.13). 37 was treated with an amine and then reacted with NaSH to produce asymmetric dinitro-arene (46). Reduction and cyclization resulted in a hybrid structure (47) that was alkylated cleanly to yield (48). Excess NaSH was combined with 37 to produce 49, which was then used to make bis(thiazole) (50).

Alkylation could be done in stages, as in the synthesis of 42, to produce asymmetric structures with two thiazolium moieties, such as 51. In summary, there is a lot of flexibility in monomer syntheses when it comes to heteroatom content, N-substituent functionality, electronic asymmetry, steric asymmetry (through judicious placement of different N-substituents), conjugation via arene linkers, solubility, and overall function.

Each synthetic route has been streamlined to be short, high-yielding, and technically simple. In the following section, we'll look at how these characteristics were used to create polymers with the desired functionality.

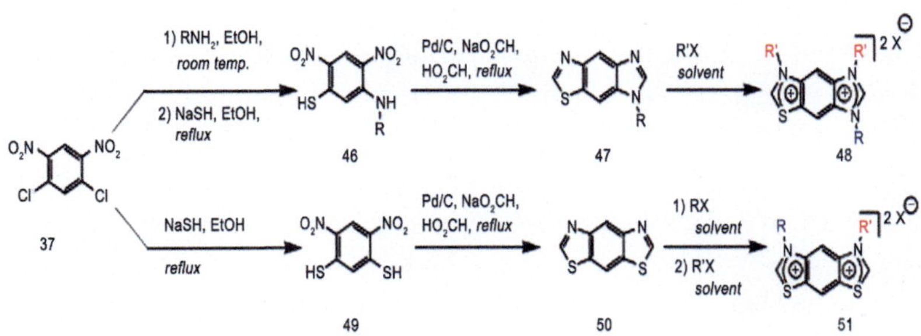

Figure 4.13. Difunctional heterocyclic carbenes as linkers.

Source: Image by springer

4.5.3. Bis(Carbene)-Based Organometallic Polymers

In the presence of acetate anion in polar solvents, all of the bis(azolium) salts presented above undergo smooth co-polymerizations with Pd(II) and Pt(II) salts (e.g. dimethyl sulfoxide [DMSO], DMF, N-methyl-pyrrolidone [NMP], CH3CN).

After precipitation into MeOH or H2O, treatment of these bis(azolium) salts with Pd(OAc)2 or PtCl2/NaOAc in polar solvents between 50 and 110 °C successfully completed the copolymerization and metal incorporation to yield the bis(carbene)-based main-chain organometallic polymers in high yield. Our initial research focused on monomers with aliphatic alkyl groups.

Poly-dispersities typical of step-growth polymerizations were discovered by GPC's molecular weight analysis. For this series, Mn was higher for Pt-containing polymers than for the structurally analogous Pd-containing systems. Notably, very high Mn (up to 1.8 × 106 Da) were obtained with this method. The electronic absorption spectra revealed a fairly narrow range of absorption maxima (308–329 nm).

Thermal stability was measured by thermal gravimetric analysis (TGA) under a nitrogen atmosphere and Td was consistently found to be between 280 and 300 °C for polymers of this structure.

The incompatibility of using metals other than Pd and Pt was a major drawback in this synthetic route. The failure to polymerize using Ni(II) salts, which have also shown nonreactivity in attempts to generate analogous small molecule Ni-NHC complexes via Herrmann's procedure, was initially of particular interest.

To verify the stability of the resultant Ni-based polymers, the free bis(carbene) was generated, which reversibly formed the corresponding homopolymer, and subsequently added anhydrous $NiCl_2$. This method provided the desired polymer as a stable macromolecule. Another alternative route utilized Lin's Ag-mediated NHC transfer reaction. Treatment of the bis(azolium) salts with Ag_2O produced thick gels when solvated which, when reacted with divalent metal halides, led to the corresponding organometallic polymers.

However, this technique produces a stoichiometric amount of metal waste and results in difficulty confirming complete trans-metallation. Procedures for small molecule NHC-Ni complexes are known that involve monomeric azolium salts with pre-dried $Ni(OAc)_2$ at high temperature under vacuum or in an ionic liquid.

Generally, the polymers in this series were poorly soluble in most common solvents (THF, CH_2Cl_2, dioxane, CH_3CN), but dissolved readily in more polar solvents such as DMF, DMSO, and NMP. Subsequent changes have involved simply using longer alkyl chains for the N-substituents (e.g., hexyl) which resulted in high molecular-weight polymers that exhibited good solubility in solvents such as THF, $CHCl_3$, CH_2Cl_2, and dioxane.

Alternatively, when increased solubility is desired in combination with relatively small N-substituents, installation of additional functionality on the arene is useful. For example, work on producing bis(azolium) salts with varying chain length alkyl groups is ongoing.

On 1,4-dialkyl benzenes, fourfold electrophilic aromatic substitution was used to produce tetra-halo or tetranitro products. The corresponding bis(azolium) salts should be obtained by reductive cyclization of the tetranitro followed by alkylation, or four-fold aryl amination of the tetra-halo followed by cyclization.

The reversibility, or "dynamicity," of bis(NHC) organometallic polymers is a key feature. The use of chain-transfer agents to modulate both polymer molecular weight and end-group structure because of the copolymerization's dynamic nature has been studied.

End-capped polymers were produced by copolymerizing Pd(OAc)2 with monofunctional benzimidazolium bromide as a CTA. Eventually, studies have focused on using reversible polymer formation to design dynamic copolymers of varying monomer structures and self-healing networks.

Another method of tailoring polymer properties is post-polymerization modification, which can be thought of as a macromolecule's modular attribute. Because the metals in bis(NHC) polymers are coordinatively unsaturated, an exogenous ligand added after polymerization was expected to bind to the metal, changing the polymer's physical and electronic properties.

The addition of PPh_3or PCy_3 to a polymer suspension in THF caused the phosphine-bound polymer to dissolve completely. ^{1}H and ^{31}P NMR spectroscopy were used to confirm the ligation.

A major breakthrough was the formation of bis(NHC)-based organometallic polymers with various transition metals of choice. Unfortunately, utilizing the initial generation of monomer scaffolds, metals with low hydrolytically stabilities as NHC complexes (e.g., Cu) were unsuitable. To solve this problem, researchers looked for a monomer that would boost the NHC ligand's affinity for the metal. This study concentrated on employing phenolic imidazoles to form an extra (ionic) bond with the metal, based on examples from similar small-molecule organometallic complexes. Dichloro-dinitrobenzene was treated with 2-aminophenol, followed by reduction-cyclization, to produce 62 using similar chemistry to that published before (Figure 4.14).

Alkylation yielded the target monomer (63) in a high percentage of the time. Polymerization was most successful with the addition of an exogenous weak base to level the mineral acid created by the reaction of the phenol with the metal halide, according to optimization studies. This study targeted transition metals such as Ni and Cu after successfully forming Pd- and Pt-containing polymers with the novel monomer design.

Small-molecule analogues are often made by fully deprotonating both the phenol and the azolium in an inert environment, then adding a soluble (ligated) metal salt. Alternatively, Hoveyda has obtained naphthalic NHC-metal complexes via Lin's NHC transfer process. There has also been no research on benzimidazolium salts functionalized with phenol substituents.

When Ni(II) and Cu(II) salts were polymerized in the presence of stoichiometric NaOAc, good yields of the respective polymers were obtained. These transition metals could be incorporated directly into high yields under ambient conditions. There was also an increase in the systems' thermal stability. All of the phenolic polymers are air and moisture stable, with Td values ranging from 340 to 362 °C according to TGA measurements in a nitrogen atmosphere.

Thermal stabilities rose by roughly 50 °C when phenoxide substituents were added to similar Pd- and Pt-based polymers carrying N-alkyl groups. The electronic absorption spectra were affected by changing the transition metal, which had a minor but noticeable effect. Each absorbs infrared light with a λmax wavelength of 287 (Ni) to 319 nm (Cu).

The improved binding affinity appeared to be customized in a way that was independent of the overall electrical nature of the system. That is, the λmax values of the phenolic Pd and Pt systems are substantially equal to those of the related N-alkyl polymers.

Directionality (head-to-tail selectivity) and mixed-metal systems are two examples of polymer design that are both noteworthy. Using the bis(carbene) strategy, researchers are aiming toward achieving these objectives. Further use of the phenol ligand, together with "left–right" desymmetrization, resulted in a designer monomer capable of forming a distinct directional polymer via bis(carbene) linker head-to-tail ordering.

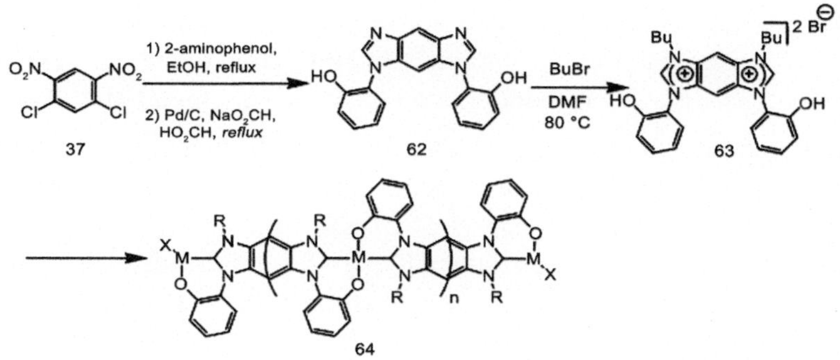

Figure 4.14. Bis(carbene)-based organometallic polymers.

Source: Image by springer

Sequential temperature-controlled SNAr reactions with an aliphatic amine followed by the addition of 2-aminophenol resulted in an asymmetric diamine with structure 43. Continuing on to the related organometallic polymer revealed that the compounds under investigation had the potential to exist.

The use of 41 to build a transition metal complex with terminal imidazole activity could lead to polymers with the structure 66, which are poised to generate an alternating bimetallic polymer with excellent control over metal placement. The wide range of transition metals that are compatible with these polymerization techniques should open up new possibilities for electrical fine-tuning.

4.6. FURTHER CONSIDERATIONS AND OUTLOOK

This study has shown that using bis(carbene) scaffolds as a monomer for organometallic polymers results in a high level of modularity. Solubility, thermal stability, metal compatibility, dynamicity, and electronic communication are just a few of the desirable properties of polymers that can be tweaked with a variety of methods.

With the addition of structurally changeable monomer topologies, the frontier of this approach to main-chain organometallic polymers will be pushed further and quicker, spanning several areas of material science. These systems' cross-linked networks are currently being explored for usage in conductive self-healing materials. Dynamic polymers as "heat-activated" catalysts are the subject of several studies. Conductivity optimization via redox matching of heterocyclic carbenes and transition metals will eventually improve the polymers' semiconducting (undoped) characteristics.

The common theme of a modular design facilitates all of these areas, allowing diversity not only in the metal included in the polymer, but also in the steric and electrical properties of the organic moieties. Due to a monomer template with numerous pathways for synthesis and, eventually, new applications, the design adjustments were made fast and efficiently.

4.7. HYPERBRANCHED POLYMERS CONTAINING TRANSITION METALS: SYNTHETIC PATHWAYS AND POTENTIAL APPLICATIONS

Over the last five decades, the incorporation of transition metals into organic monomers and polymers has been widely investigated due to the promise electrical, magnetic, optical, sensing, and catalytic capabilities that these organometallic compounds possess.

Organometallic polymers have found applications in the coating, pharmaceutical, and aerospace industries due to their fascinating features that are typically unavailable to their pure organic progenitors.

Many of these studies focused on the synthesis of linear polymers with transition metals integrated into the main chain or attached as pendant groups at the side chains, but the preparation and study of highly branched three-dimensional (3-D) macromolecular architectures—such as dendrimers and hyperbranched polymers containing organometallic complexes—has only recently received more attention.

Metal centers have been shown to act as cores, simulating artificial models of biological systems such as metalloenzymes, as well as connectors, branching points, and terminal (surface) units distributed throughout the structure with potential applications in sensors, catalysts, and light-harvesting antennas, depending on their position.

Despite their structural beauty, dendrimer synthesis necessitates a careful approach requiring multistep reaction and purification methods in order to produce the different tree-like generations, limiting their potential applicability to academic interests only in many cases. Furthermore, recent developments in mass spectrometric techniques have shown portrayals of dentrimers that are extremely idealized, despite the fact that real samples do include flaws and structural defects. Despite their random and polydisperse architectures, "imperfect" hyperbranched polymers often exhibit similar, if not identical, molecular characteristics to their "perfect" dentrimer congeners. Nonetheless, single-step polymerization processes can easily prepare them, allowing for large-scale manufacture and thereby expanding their potential uses and applications. The synthesis of hyperbranched conjugated organic and organometallic polymers is of great interest to research groups. By polymerizing metal-containing monomers or post-functionalizing the hyperbranched scaffolding with organometallic complexes, researchers have established various synthetic methods toward high-metal-loaded materials. AB_n

4.7.1. Research and Discussion

Since Kim and Webster's seminal study, hyperbranched polymers have been the subject of ongoing research due to their tree-like molecular architectures. Three major components can be seen throughout the structure (Figure 4.15): dendritic or branching units (D), linear units (L), and terminal units (T).

For the manufacture of pure organic hyperbranched polymers, many synthetic techniques have been used. Self-condensation of AB_n -type monomers with $n \geq 2$ is the most widely used method. This sort of polymerization can be done simultaneously, by slowly adding the monomer, or even in the presence of a Bf ($f \geq 3$) core molecule, allowing for structural control over the emerging polymer.

Co-polymerizations of A2 monomers with Bn comonomers ($n \geq 3$) are another option.

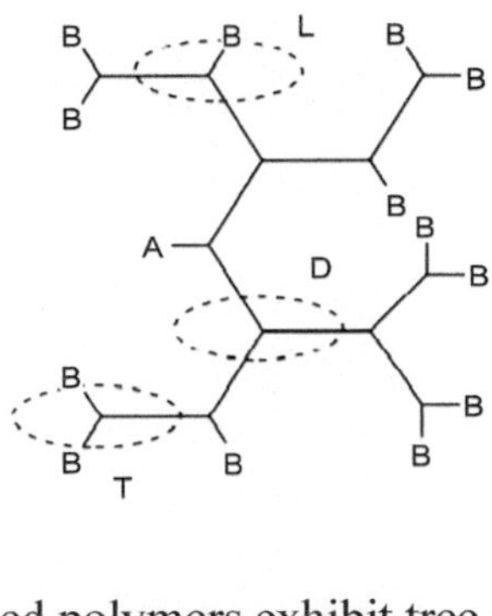

Figure 4.15. Hyperbranched polymers exhibit tree-like molecular structures.

Source: Image by springer

The stoichiometric requirements between the pairs of functional comonomers, as well as the potential for gelation, are significant disadvantages. Frechet reported the synthesis of hyperbranched polymers by self-condensing vinyl polymerization (SCVP), which has recently been applied to a variety of other living/controlled polymerizations, including nitroxide-mediated radical polymerization, atom transfer radical polymerization, group transfer polymerization, and ring-opening polymerization.

4.8. SYNTHETIC PATHWAYS

All of the recognized methods for knitting metal containing monomers together could be used to directly create organometallic hyperbranched polymers, providing they are stable under the applied polymerization conditions and do not interfere with the reaction mechanism. Organometallic complexes can also be used to functionalize suitable pure organic hyperbranched polymers. Other study organizations have used both methodologies in the past.

4.8.1. Incorporation of Transition Metals through the Building Block

Although there are numerous methods for creating hyperbranched polymers, there are few examples of metal-containing compounds that are directly produced from organometallic monomers. Reinhoudt and colleagues described the self-assembly of hyperbranched polymers using an AB2-type monomer made up of organopalladium complexes sandwiched between SCS pincer ligands and labile acetonitrile molecules.

As demonstrated by atomic force microscopy (AFM) and transmission electron microscopy (TEM), ligand exchange by solvent elimination leads to reversible spherical assemblies. The size of the spheres can be controlled by adjusting the counter anions and manipulating the substituents on the pincer ligand.

Linear analogues, on the other hand, did not display any globular formations, supporting the need for branching units.

Lewis et al. used the A2 + B3 technique to create hyperbranched organometallic polymers by treating Pt(PBu3)Cl2 with 1,3,5-triethynylbenzene in a 3:2 molar ratio. However, the resultant compound was insoluble in typical organic solvents, and the only way to stop the cross-linking processes was to add too much p-1,4-diethynylbenzene (triyne:diyne = 1:50).

Takahashi and colleagues used a self-polycondensation reaction to create a formally comparable hyperbranched polymer from the AB2-type analogue of Lewis' monomers in an alternate way. The resulting organometallic polymer was soluble in common organic solvents and could be identified via spectral and gel permeation chromatography (GPC).

Abd-El-Aziz and colleagues recently reported the development of hyperbranched poly(aryl-ethers) (4,5) and poly(aryl-thioethers) (6) containing cyclo-penta-dienyl-iron moieties by nucleophilic substitution of A2 + B3 type monomers (Figure 4.16).

Standard spectroscopic analytical techniques revealed that the polymers had low viscosities and that the organometallic complexes were stable up to 230 °C, as determined by thermal gravimetric analysis (TGA).

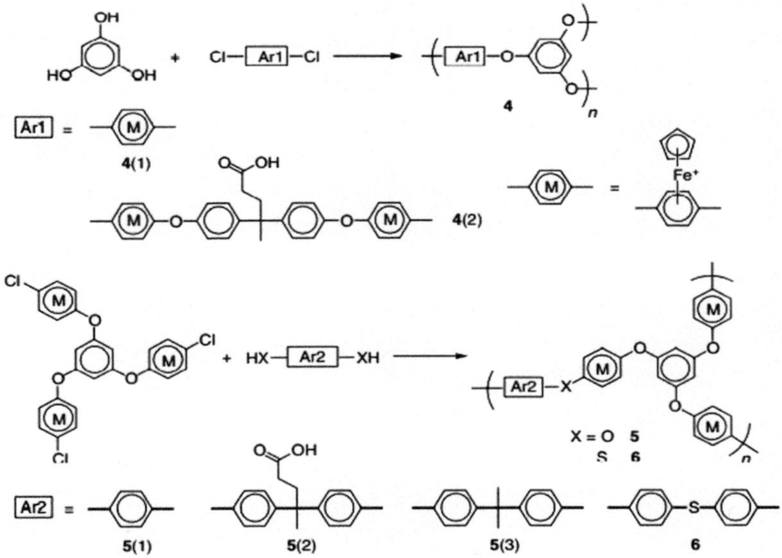

Figure 4.16. Synthesis of cyclo-penta-dienyl-iron-containing polymers via A_2 + B_3 method.

Source: Image by springer

Ferrocene is an appealing building element for highly branched materials preparation. Galloway and Rauchfuss described the desulfurization-induced ring-opening polymerization of high-molecular-weight poly (ferrocenylenepersulfides) (ROP). Similarly, this group used ferrocene as a metal-containing building block and used salt-eliminative poly-coupling of 1,1-dilithioferrocene with alkyl-trichloro-silanes to create hyperbranched poly(ferrocenylsilanes). From methyl to n-dodecyl-substituted polymers, increasing spacer length enhanced solubility and molecular weight.

The polymers have robust skeleton structures with prolonged conjugations, and their absorption spectra tail into the infrared region (> 700 nm), according to spectroscopic investigations. Other group 14 and 15 elements, such as germanium, phosphor, and antimony, have recently been included in this approach.

However, because most of the polymers had low solubility, a comprehensive structural study was impossible. Nonetheless, by heating in an inert gas atmosphere, all of the hyperbranched polymers acted as effective precursors for the formation of metal-containing ceramics. The pyrolytic yields were found to be higher than their linear analogs in general.

Ceramics containing mainly -Fe nanoparticles were produced by calcining Si-containing polymers at 1,000 °C under nitrogen, but those containing Ge- and Sb-containing polymers were totally changed into corresponding iron-alloys. Iron phosphide diffraction patterns were visible in the ceramics made from P-containing polymers.

When the pyrolysis of the methyl-substituted hyperbranched poly(ferrocenylsilane) 11(1) was carried out at a higher temperature of 1,200 °C under argon, bigger iron silicide nanocrystals were produced. With magnetic saturations (Ms) up to 51 emu/g and near-zero remanence and coercivity, this ceramic was extremely magnetizable.

Ongoing research has recently resulted in the formation of a new methodology for the synthesis of hyperbranched poly(aroyl-arylene)s using amine-catalyzed regioselective poly-cyclotrimerization of bis(aroyl-acetylene)s containing ferrocene moieties (Figure 4.17).

Incorporation of the ferrocene motif was achieved in yields up to ~70% (Mw 9 100, Mw/Mn = 2.8-3.1) by either homo-poly-cyclo-trimerization of diyne 12 or co-poly-cyclo-trimerization of 14 with monoyne 15.

Furthermore, the two synthetic techniques, homo- and co-poly-cyclotrimerization, allow for molecular structural tailoring, with the ferrocene building blocks dispersed evenly throughout the hyperbranched polymer or mostly on the outside as terminal units. Numerous benzophenone and tri-aroyl-benzene functionalities are present in the polymers, which are known to readily cross-link when exposed to ultraviolet (UV) radiation or other high-energy sources.

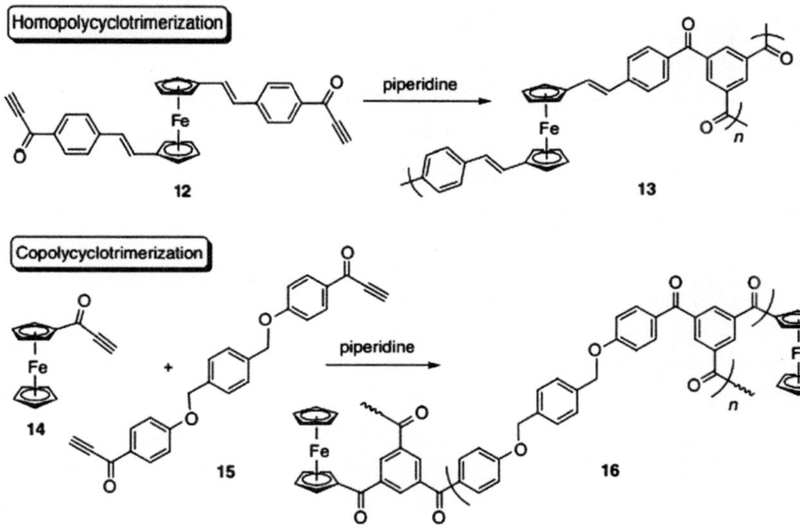

Figure 4.17. 1,3,5-Regioselective homo- and co-poly-cyclotrimerization of ferrocene-containing aroyl-acetylenes.

Source: Image by springer

Ceramization of 16 silicon wafers at 1,000 °C for 1 hour in a tube furnace under nitrogen steam produced a ceramic pattern with exceptional form preservation compared to their polymer antecedents.

Under increasing magnification, the ceramic pattern displayed a morphological shift from a homogeneous thin-film to congeries of microscopic ceramic clusters. Fe and Fe2O3 nanoparticles embedded in a carbon matrix make up the composition of these ceramic patterns.

4.8.2. Incorporation of Transition Metals through Post-Functionalization

Organometallic polymers can be made directly from their metal-containing monomer building blocks or through post-functionalization reactions, in which suitable chelating groups within the molecular architecture act as macro-ligands for metal complexes and nanoparticles.

The hyperbranched scaffoldings could be used as homogeneous nanoreactors for the inclusion of catalytically active metal species with this concept in mind. Such macro-catalysts might also be easily recovered and utilized in another reaction cycle using precipitation or filtration procedures.

Metal complexes can be injected in the core, on the surface (along the numerous terminal units), or across the entire hyperbranched structure, depending on the position of the chelating functional groups.

Frey and van Koten's groups pioneered this subject by functionalizing hyperbranched carbo-silane polymers with palladium complexes and used them as homogenous organometallic catalysts for a conventional aldol condensation reaction.

The reactivities of the hyperbranched polymer-supported metal catalysts were remarkably similar to related den-trimers, showing that structural perfection isn't necessarily essential. This nano-capsules concept was recently expanded to amphiphilic hyperbranched polyglycerols, which fixed pincer-platinum(II) complexes and palladium(II) salts selectively within the hydrophilic core.

The catalytic activity of the Pt-containing nano capsules was examined in a double Michael addition process, and it was shown to be lower than that of the respective unsupported catalyst, probably due to the catalyst's limited accessibility in the nano capsule's interior.

Furthermore, the asymmetric model addition of methyl vinyl ketone and -cyano-propionate created no enantiomeric excess in the optically active hyperbranched analogue, indicating that the chiral nano capsule backbone had no effect on the resultant product. The Pd salts might be converted to metallic nanoparticles, which the hyperbranched scaffold could help to stabilize. The resultant Pd-colloids were tested as homogeneous catalysts for cyclohexene hydrogenation and found to be more active than commercially available Pd/activated charcoal catalysts.

A hyperbranched polymer with a structure identical to 19, but with 1,2-dimethylimidazolium end groups instead of palmitoyl end groups, was functionalized by counterion exchange with monosulfonated triphenylphosphine. The immobilization of [Rh(acac)(CO)2] onto the hyperbranched surface resulted in polymer-bound complexes with considerable activity in the methanol hydroformylation of 1-hexene.

Other researchers reported the formation of metal nanoparticles such as Ag, Au, Cu, Pt, and Pd stabilized by various hyperbranched polymers such as poly(ethylene-imides), poly(amidoamines) (structurally similar to PAMAM den-trimers), poly(amine-esters), and aromatic poly(amides) and evaluated their activity towards various chemical reactions.

By using electrostatic forces, Bai and colleagues were able to insert CdS nanoparticles into the cavities of a hyperbranched conjugated poly(phenylenevinylenes) (PPV) with various alkoxy side chains. The findings reveal that hyperbranched polymers were able to efficiently transfer energy to semiconductor nanoparticles, while reducing self-aggregation. Water-solubilizing sulfonic groups were used to improve nanoparticle dispersion.

During this study, a totally soluble high molecular weight conjugated hyperbranched polyynes (24), whose triple bond scaffold worked well as a macro-ligand for cobalt carbonyls was created (Figure 4.18). The cobalt-clusterized hyperbranched polyynes (25) were successfully changed into advanced ceramics with remarkable soft-ferromagnetic characteristics and magnetic saturations of up to ~118 emu/g.

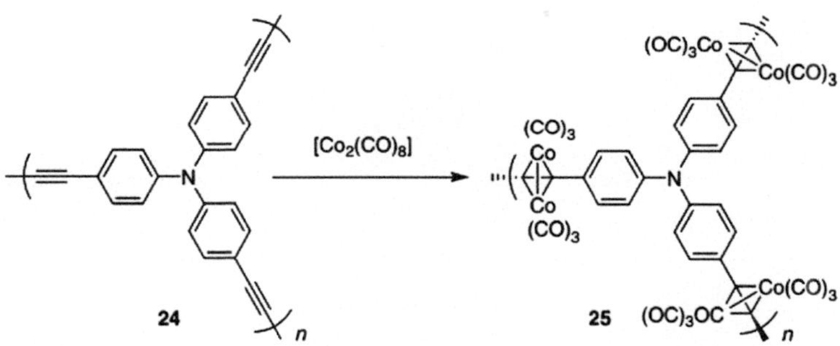

Figure 4.18. Synthesis of cobalt-containing hyperbranched polyynes.

Source: Image by springer

Unfortunately, once precipitated in a weak solvent, the high metal-loaded polyyne becomes partly soluble or even insoluble due to the development of supramolecular aggregates.

By spin-coating the freshly generated solutions of the organometallic polymers onto silicon wafers, thin films of good quality (thickness: 1000 nm, as confirmed by ellipsometry) could still be formed. The films were photobleached after being photopatterned with a Cu-negative mask, and features in the size range of 10 to 100 m were easily obtained.

Wavelength-dependent refractive index studies of unexposed and exposed thin films of 25 yielded surprising results. While the metallified polymer, like its non-metallated parent (24), had very high refractive indexes (n = 1.8131.714) in the 600 to 1600 nm spectral region, the refractive index dropped significantly (n = 1.7771.667) after UV irradiation, possibly due to the decomposition of the organometallic moieties.

Photonic applications may benefit from materials with such a high refractive index change: It might, for example, be used as a photorefractive material in holographic systems or as optical coatings with a high refractive index.

The formation of carbon nanotubes (CNTs) through chemical vapor deposition is known to be catalyzed by cobalt complexes, among other metals such as iron and nickel (CVD).

Spin-coated films of the organometallic polymer were successfully investigated to operate as catalyst and arrays of CNT bundles were created thanks to the thermal stability of the hyperbranched polyyne backbone. This preliminary finding points to a possible application in the field of patternable, custom-made catalysts.

4.9. POLYMERIC ORGANOTIN FIBERS

The inability of most metal-containing polymers to form fibers is due to a lack of sufficient polymer solubility and, as a result, fiber formation. It's much more unusual for polymers to generate fibers on their own. More than three decades ago, we looked into this trend for a variety of Group IVB polyesters. In a nutshell, the polymers were created via the interfacial polymerization technique.

The product is precipitated from the reaction, collected on filter paper in a Buchner filter with suction, rinsed with the organic solvent and water to remove unreacted components and salts, and then removed from the filter paper in a glass Petri dish with acetone. Allow time for the product to dry. Many of these compounds created fibers when scraped from glass Petri dishes.

Visual and microscopic examinations revealed no fibers in general. Fibers formed spontaneously as the product was collected from the dish. Fibers were present in certain cases as the liquid evaporated. The fiber was commonly created by scraping the polymer with a flat-ended steel spatula. The mechanical agitation appears to be adequate to stimulate fiber production. The majority of fiber-forming structures, such as those formed from terephthalic acid, have rigid backbones. Others, such as itaconic acid, have semi-rigid backbone architectures (methylene succinic acid). But others, such as those derived from azelatic acid, have very flexible backbones (non-anediolic acid). Although the structures were all drawn for M = Ti, fibers were also produced by Zr and Hf products.

These fibers' architectures and physical qualities were investigated and compared to nonfibrous components of the product. Differential scanning calorimetry revealed phase transitions in both air and nitrogen at approximately the same locations and magnitudes.

The degradation was the same. While DSC detected transitions in the 90 to 250 °C range, weight loss was observed for several of the products up to 1,000 °C, with nitrogen exhibiting less weight loss than air. The infrared spectra of the materials before and after heating were very comparable.

Finally, in the range of 3,000 to 200 cm^{-1}, there were no discernible variations in the infrared spectra of fibrous and nonfibrous parts of the same polymer. For the terephthalic acid-derived product, a sharp band 3,130 cm^{-1} was frequently associated with the Cp and terphthalate moieties in nonfibrous products.

In the fibrous product, this peak was substantially obscured by bands ranging from 3,480 to 3,100 cm^{-1}. When comparing fibrous and nonfibrous products, these bands were more pronounced for fibrous items. Increased hydrogen bonding could explain the increased intensity of these bands.

The fibers were flexible, and some of them remained flexible up to 500 degrees Celsius. For more than 30 years, the fibers maintained their original elasticity and other qualities. We've only detected fiber production in a few items since then, and we've only mentioned it in passing when discussing the synthesis of novel polymers.

A number of fiber-forming products have recently been discovered, some of which are produced from acyclovir and other metallocene dichlorides such as vanadocene and niobocene. This section describes the formation of fibers from a simple organotin polyether and various organotin derivatives.

As a side aside, the synthesis process may be important for fiber production. When using "high-energy" Lewis acids, such as acid chlorides, the interfacial polymerization system is thought to provide not only a quick alternative technique for polymer synthesis, but also some orientation to the developing polymer chains.

This orientation is most likely the product of polymerization taking place in a two-dimensional, layered environment comparable to lignin. This results in a layered, two-dimensional structure for lignin. When other polymers were produced within cavities that resembled two-dimensional templates, similar structures were discovered. The majority of self-assembly polymerizing systems have similar restrictions. As a result, this inclination is not unusual.

4.9.1. Organotin Poly-Ethers

A number of organotin condensation polymers were created for a variety of reasons, one of which was to investigate their biological activity. Organotin polymers, as well as their application as anticancer drugs, were recently discussed. Bu > Pr > Et > Me, with the methyl, octyl, and lauryl groups being largely inert, is the order of capacity to limit cell growth with respect to the alkyl chain on the organotin. A number of organotin poly-ethers of the general form were made a long time ago.

Some of these compounds were studied for their ability to suppress Balb 3T3 cell growth as a measure of their anticancer potential. Only dibutyltin products will be discussed here. The 1,6-hexanediol product has a GI50 of 5 g/ml (growth inhibition of 50%).

The GI50 for the dibutyltin dichloride product of 1,4-butanediol was 0.25 g/ml. The GI50 for the 1,4-butenediol product was 0.025 g/ml, the lowest GI50 for the organotin polymers so far. By comparison, cisplatin, the most extensively used anticancer medication, has a GI50 of 0.4 g/ml.

These findings led to the discovery of two structural windows that needed to be studied further. First, activity increased as the distance between the oxygen atoms shrank. Second, unsaturation—the presence of bonds—could play a role in the organotin poly-ethers' ability to hinder cell growth. The compound of dibutyltin dichloride and 1,4-butynediol was produced to investigate these windows.

1,4-butynediol and dibutyltin dichloride were acquired from Aldrich. The interfacial polycondensation approach was used to carry out the reactions.

Briefly, a 30 ml aqueous solution comprising the diol (0.00300 M) and sodium hydroxide (0.0060 M) was transferred to a one quart Kimax emulsifying jar installed on top of a Waring Blender (model 1120; no load speed of about 18,000 rpm; reactions carried out at about 25 °C). A hexane solution (30 ml) containing dibutyltin dichloride (0.00300 M) was rapidly introduced through a hole in the jar lid using a powder funnel (over 3–4 seconds).

For 15 seconds, the resultant solution was mixed. To eliminate unreacted elements and undesirable by-products, the precipitate was vacuum filtered and washed numerous times with deionized water and chloroform. The solid was rinsed and dried at room temperature in a glass Petri plate.

Researchers discovered a striking resemblance in the structures that are prone to forming these fibers: stiff structures are more likely to create fibers. Several of the poly-ethers produced in this investigation did not generate fibers after being recovered from the Petri dish. Ethylene glycol, 1,6-hexanediol, 1,4-hexanediol, and 1,4-butenediol were among the structures studied.

Fibers were detected in the similar polyether prepared with 1,4-butynediol. To be consistent with the previous observation, the 1,4-butynediol product should be the stiffest of the poly-ethers synthesized in this sequence. Because of the delicate balance necessary in producing these fibers, researchers performed the reaction numerous times with variable outcomes.

Some fibers were created in each case, but in varying amounts. The result was polymeric, with a molecular weight in HMPA of 12,0000, equivalent to an average chain length of 380, as determined by light scattering photometry. For 5 weeks, the molecular weight was measured weekly and did not change. As a result, the polymer can be kept in solution for at least a month.

The ability of the product to inhibit Balb 3T3 cells was examined. They had a GI50 value of 0.05 g/ml, according to preliminary studies. Cisplatin, the most extensively used anticancer medicine, has a GI50 value of 0.50 g/m, which is nearly 10 times higher than the concentration found in the dibutyltin/2-butyne-1,4-diol product. This was in line with the theory that a structural window with a small number of carbons between oxygen and the presence of unsaturation are advantageous in the development of anticancer medicines that limit cell growth at low doses.

4.9.2. Application

The most obvious field of application is as composite fibers. Metallic whiskers are used as high-strength fibers in composites in a variety of ways. These composites are among the most powerful ever created. Fiber mechanical and electrical qualities must also be investigated. It's likely that the fibers are semiconductors when they come close to conductors, allowing them to function as directional electrical lines.

4.10. CONCLUSION

In this study, the most recent developments in the field of hyperbranched organometallic polymers in this contribution have been highlighted. The synthesis of hyperbranched polymers from metal-containing building blocks and the post-functionalization of hyperbranched structures have been examined, with fascinating inorganic–organic hybrid materials emerging.

Existing evidence suggests that structural perfection, such as that observed in den-trimers, isn't necessarily crucial. Future applications in the fields of homogenous reusable catalysis, various types of adhesives and coatings, photoresists, and processable precursors for advanced materials, in particular, are already on the horizon, advancing this new but promising research area.

REFERENCES

1. Boydston, A. and Bielawski, C., 2008. Synthetic Versatility and Structural Modularity in Organometallic Polymers. *Inorganic and Organometallic Macromolecules*, [online] pp.1-20. Available at: <https://link.springer.com/chapter/10.1007/978-0-387-72947-3_1> [Accessed 29 June 2022].
2. Boydston, A., Williams, K. and Bielawski, C., 2005. A Modular Approach to Main-Chain Organometallic Polymers. *Journal of the American Chemical Society*, [online] 127(36), pp.12496-12497. Available at: <https://pubmed.ncbi.nlm.nih.gov/16144390/> [Accessed 29 June 2022].
3. Dixon, N., Lawrance, G., Lay, P. and Sargeson, A., 1984. Synthetically versatile (trifluoromethanesulfonato)metal amine complexes. *Inorganic Chemistry*, [online] 23(19), pp.2940-2947. Available at: <https://pubs.acs.org/doi/abs/10.1021/ic00187a009> [Accessed 29 June 2022].
4. Ortiz, A. and Sansinenea, E., 2007. The synthetic versatility of oxazolidinethiones. *Journal of Sulfur Chemistry*, [online] 28(2), pp.109-147. Available at: <https://www.tandfonline.com/doi/abs/10.1080/17415990601139699?journalCode=gsrp20> [Accessed 29 June 2022].
5. Serrano-Ruiz, M., Scalambra, F. and Romerosa, A., 2013. Organometallic Polymers. *Advances in Organometallic Chemistry and Catalysis*, [online] pp.379-405. Available at: <https://onlinelibrary.wiley.com/doi/10.1002/9781118742952.ch29> [Accessed 29 June 2022].

CHAPTER 5

Plant Macromolecules as Biomaterials

CONTENTS

Skin is the body's biggest organ, with a reasonably strong arrangement that serves as a natural screen against physical, chemical, and bacterial harm. When this pattern is disrupted by acute lesions, the skin goes through a remarkable healing process that results in wound closure and the formation of scars.

To maintain the normal control of the skin, this complicated system relies on various cell types and mediators interacting. In instances such as diabetes or vascular lesions, faulty skin frequently persists, causing loss of integrity and, in certain cases, overlaying of structural layers, as well as delayed skin tissue healing.

5.1. INTRODUCTION

The large molecular weight of plant macromolecules distinguishes them from all other components. This can range from 10,000 to over 1,000,000, although the molecular weight of other plant metabolites is rarely greater than 1,000. Chemically, macromolecules are lengthy chains of tiny structural units or 'building blocks that are covalently bonded in a variety of ways.

In the first instance, chemical characterization is dependent on detecting these smaller components. Proteins, for example, are long chains of amino acids (up to twenty distinct types) connected by peptide (-CO-NH-) connections. Polysaccharides are formed similarly by the union of simple sugar units, such as glucose, linked by ether (-0-) linkages.

Nucleic acids, on the other hand, are more complicated, with three structural units: purine and pyrimidine bases, pentose sugars, and phosphate groups. Proteins, polysaccharides, and nucleic acids are therefore the three major groups of macromolecules present in plants. However, mixed polymers, such as glycoproteins, which include both sugars and amino acids in covalent connection, are also known.

While the polymers listed above all have an ordered structure, plants contain random polymers generated by oxidative, non-enzymatic polymerization of simple phenolic components. There are three types of these: lignins, which are generated from phenylpropanoid units, condensed tannins, which are derived from flavonoid units, and plant melanins produced from catechol polymerization.

Thus, DNA and RNA play roles in the storage and transcription of genetic information, proteins serve as catalysts in enzymatic activities, polysaccharides serve as a kind of energy storage, and so on. However, several plant macromolecules have a specialized role connected to plant development.

For example, the production of lignin, a polymer that, in conjunction with cellulose, creates a complex matrix in the cell wall, allows the plant to build a solid stem or trunk. Plant macromolecule separation and identification methods differ from those used for low molecular weight components in several ways.

Solubilizing polymers, for example, is frequently challenging, and additional techniques may be required. In a simple scenario, macromolecular components can be dissolved by homogenizing plant tissue with salt solution and subsequently precipitated by adjusting the extract's pH.

Because the concentration of various macromolecules (especially nucleic acids) in plant tissues is low in general, it may be useful to break the plant cells and separate the different organelles by centrifugation prior to isolation. The mitochondrial fraction will be high in enzymic protein, whereas the nuclear fraction will be high in nucleic acid.

Because of the structural complexity of macromolecules in live cells, any method of separation will almost always result in some degradation or artifact production unless particular care is followed. Enzymic protein is particularly prone to breakdown and loss of function during isolation techniques, therefore it is critical to prevent pH and temperature extremes.

Again, due to their strong relationship with carbohydrates, lignins are difficult to get pure. The study of macromolecules, whether from plant or animal tissues, is crucial to understanding molecular biology, and much has been published about the methodologies necessary for their characterization elsewhere.

There are also numerous thorough texts on the methods of separating proteins and nucleic acids, primarily from animal tissue but also from plant tissue. It will be impossible to address every facet of this large subject in the space given here.

This description will focus on the difficulties connected with isolating macromolecules from plant tissues rather than animal tissues. Plant polymer comparative features will also be examined in depth.

Human bodily tissues have a variety of physical properties, including stiffness and porosity. Tissue engineering has benefited greatly from a multidisciplinary interface across topics such as cell biology, biotechnology, mechanics, materials science, bioengineering, and clinical research.

Many wound dressings have been designed with the goal of restoring and improving tissue function by creating new biocompatible replacements or repairing these tissues. However, these materials are relatively expensive, which may limit their broad usage.

According to this viewpoint, an ideal wound dressing would preserve a microenvironment in the wounded bed and would direct particular healing characteristics for each type of wound or illness that affects the patient being treated.

The wrapping also should maintain hydration levels, which acts as a barrier against infections and provides thermal insulation for the wound. As a result, recent biomedical technology investments in the formation of scaffolds capable of mimicking the natural environment for skin growth and regeneration after injury.

Polysaccharide-based biomaterials are making a name for themselves in the field of tissue engineering, primarily as hydrogels for the successful treatment of wounds and skin burns. They are classified as neutral (glucans, dextran, and cellulose, for example), acids (hyaluronic acid), basic (chitosan), or sulfated polysaccharides (heparin and chondroitin).

Chitosan, hyaluronic acid, and alginate are the most common and naturally occurring polysaccharide-derived biomaterials. These polysaccharides are further classified as homopolysaccharides (glucans, cellulose, dextran, and chitosan) and heteropolysaccharides (alginates, agarose, carrageenan, pectin, galactomannans, and xyloglucans).

All have unique physicochemical features as well as high biocompatibility and biodegradability, making them useful in biomedical domains. We are particularly interested in plant macromolecules such as carbohydrates and proteins (particularly hemicelluloses and lectins) as biomolecules for wound healing.

Carbohydrates are perhaps the most numerous and significant biomacromolecules in Nature. Carbohydrates, with the exception of energy-related molecules, may be separated into two categories: carbohydrates as matter and carbohydrates as information.

Carbohydrates are plentiful as materials in the extracellular matrix of mammals and cell walls of diverse plants, bacteria, fungi, and other organisms, acting as scaffolds. Some common polysaccharides are characterized as biocompatible materials with adjustable stiffness and functionality, resulting in polymeric biomaterials that are widely employed in drug delivery, tissue engineering, and other applications. Carbohydrates are commonly referred to as glycans in glycoproteins, glycolipids, and proteoglycans that can attach to proteins or other carbohydrates and thereby mediate cell-cell and cell-matrix interactions.

These glycans might be reduced to synthetic glycopolymers, glycolipids, and glycoproteins by polymerization, multistep synthesis, or a semisynthetic method. Carbohydrates' information role may be proved not just as targeting reagents, but also as immunological antigens and adjuvants. These are also covered in this section because they are always found in a macromolecular composition. In this section, carbohydrate-based macromolecular biomaterials are summarized since 2010, with an emphasis on basic knowledge to drive biomaterial rational design.

Natural occurring polysaccharides, naturally generated synthetic polysaccharides, glycopolymers/glycodendrimers, supramolecular glycopolymers, and synthetic glycolipids/glycoproteins will be examined in terms of their resources and chemical structures.

The multiscale structure-function links in numerous significant application domains, such as delivery systems, tissue engineering, and immunology, will be discussed in depth. This study will be useful in the development of carbohydrate-based macromolecular biomaterials and will help to bridge the gap between carbs as matter and carbohydrates as information in order to encourage novel biomaterial design in the coming years.

5.2. PLANT POLYSACCHARIDES

Plant polysaccharides are macromolecule complexes made up of several similar or distinct monosaccharides linked together by α- or β-glycosidic linkages. They are found in plants, such as starch, cellulose, and pectin. Because plant polysaccharides are distributed widely, the molecular composition and molecular weight of polysaccharides from various species vary.

Polysaccharides derived from plants have received a great deal of interest in recent decades due to their substantial bioactivities and suitability for therapeutic applications. Previous research has also found that plant polysaccharides are not poisonous and have no negative side effects.

Plant polysaccharides, as an active portion of antihyperglycemic plants, play an essential role in the treatment of diabetes management.

Tea polysaccharides, *Lycium barbarum* polysaccharides, maize silk polysaccharides, Atractylodes polysaccharides, yam polysaccharides, and other plant polysaccharides have been shown to lower blood sugar levels.

Because most plant polysaccharides are structural elements of cell walls, the extraction process is determined by the structure of the cell wall. After pretreatment, tea polysaccharides are extracted using solvent extraction (water, acidic, or dilute alkaline solution) to remove fat, monosaccharides, oligosaccharides, alkaloids, and polyphenols.

Recently, novel techniques for extracting tea polysaccharides have been developed, including freeze-thaw, microwave, ultrasonic, enzymatic, and high-pressure procedures. Alcohol precipitation, dialysis, ultrafiltration, resin or DEAE chromatography, Sephadex chromatography, and CTAB (hexadecyl trimethyl ammonium bromide) precipitation are all recognized isolation procedures. Tea polysaccharide yield and bioactivity vary according to the extraction and isolation method utilized.

Many investigations on the extraction and isolation of tea polysaccharides have been conducted. In one study, green tea crude polysaccharides were extracted and isolated using three different extraction methods, including ethanol precipitation, ultra-filtration technology, and CTAB precipitation, and the compositions and biological activities of the resulting polysaccharides were studied and compared.

The results revealed that ultrafiltration produced 23.5 percent and 37.1 percent more active polysaccharides than the other two techniques, correspondingly. Li et al. researched microwave-assisted tea polysaccharide extraction and determined the best extraction parameters and rate using orthogonal experiments based on a single factor experiment:

Microwave power 500 W, (W/V) 1:35, extraction temperature 50 °C, extraction period 5 min, optimum product yield 1.99 mg/g Enzymes were used in an extraction procedure with ideal circumstances of temperature 100 °C, W/C 1:26.8, ethyl alcohol concentration 90 percent, and product yield 7.86 mg/g. Under these extraction conditions, the yield of tea polysaccharides was 25.6 percent greater than without enzyme technology.

Our research created a method for extracting components from tea in an efficient and complete manner utilizing resins. The very same green tea source yielded three distinct products: tea polysaccharides, tea polyphenols, and caffeine. The extraction and isolation processes have an impact on the product's purity and bioactivity.

Plant polysaccharides are biodegradable polymers with stable structures, and various biological, physicochemical, and hydrophilicity/viscosity qualities that influence the fluid system's rheological properties. Starch is a plentiful polysaccharide biopolymer that has potential applications in medication administration and wound treatments.

The primary components of starch (amylopectin and amylose) can be physically/chemically modified to be used in medical applications. Wet electrospinning is a green technology that uses sodium palmitate to create pure starch-based nanofibers with great water stability at ambient temperature.

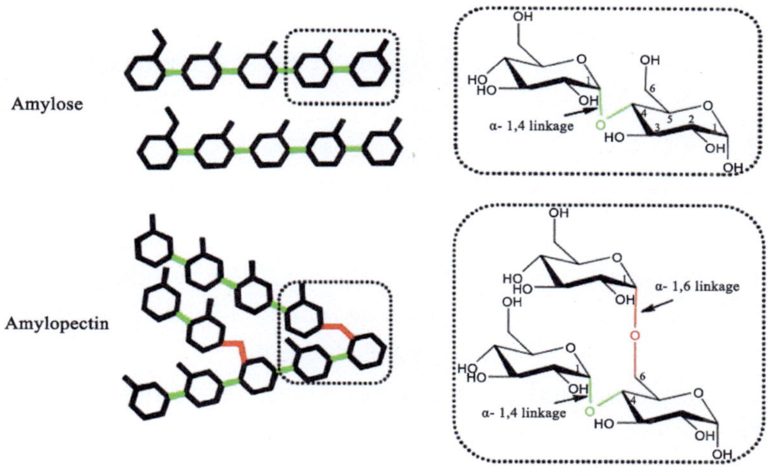

Figure 5.1. Components of Starch.

Source: Image by Wikimedia Commons

The present focus of research is on producing crosslinked electrospun nanofibers, such as PVA/starch/chitosan nanofibrous mats in wound dressings. The improved coaxial electrospinning results in a core-shell starch-hyaluronic acid/polyurethane-based electrospun nanofibers patch that is biocompatible, biodegradable, confers surface hydrophilicity and increases mechanical durability.

Nanoscale cellulose fibers are appealing due to their larger surface area when compared to their microscale counterparts. Many cellulose derivatives, including carboxymethyl cellulose (CMC), cellulose acetate, -cellulose, and ethyl cellulose, produce various electrospun nanofibers in wound dressings.

Unlike drug-free -cellulose nanofibers, antibiotic-loaded -cellulose nanofibers reduce wound size while providing the desired bioactivity. Electrospun nanofiber scaffolds based on pectin may serve as superior wound dressings with high antibacterial activity.

They absorb more exudate in less time than electrospun nanofiber patches based on alginate or chitosan. Polysaccharides from pectin are complicated compounds with uneven carbohydrate chains. Because of their mild hydrophilicity, pectins operate as an exudate-absorbing component in hydro colloidal wound dressings.

Gums, a major polysaccharide group, are a unique source of electrospun nanofiber biopolymers with many medicinal uses. Electrospun nanofiber scaffolds including gum Arabic, gum karaya, gum guar, Ajivas gum, and gum tragacanth, for example, are utilized in medical applications.

5.3. PLANT MACROMOLECULES AS BIOMATERIALS FOR WOUND HEALING

To sustain skin function, wound healing is a crucial, complicated, and highly coordinated process. A plethora of molecular and cellular processes are engaged immediately after damage to stop blood loss, remove bacteria and foreign materials, and recompose wounded tissue.

These biological and physiological stages in wound repair may be classified as follows: hemostasis, inflammatory response, cell proliferation and formation of the extracellular matrix (ECM), and the latter period, known as remodeling. These stages are not mutually exclusive; rather, they overlap throughout time.

Platelets circulate near the vascular walls in physiological conditions and are activated when the continuity of the endothelial layer is ruptured and the underlying subendothelial matrix is exposed, initiating the first stage of tissue repair, which is characterized by hemostasis and the formation of a matrix in the wound bed.

This matrix is the consequence of circulating platelet adhesion and aggregation to the components of the underlying ECM. Extrinsic and intrinsic coagulation pathways are activated by damaged tissue and aggregated platelets to stabilize the fibrin platelet clot.

This entire process serves as a framework for the migration and proliferation of additional wound-healing cells, as well as a reservoir for cytokines and growth factors. After then, the inflammatory phase begins as an innate immune response to enhance the clearance of cellular and extracellular waste as well as harmful bacteria.

Platelets and leukocytes both produce inflammatory cytokines, creating a chemotactic gradient that attracts more leukocytes and amplifies the inflammatory process. Interleukin-1, IL-1, IL-6, IL-8, tumor necrosis factor (TNF), platelet-derived growth factor (PDGF), and transforming growth factor- are examples of inflammatory factors.

Clearly, PDGF is critical early in the chemotaxis of neutrophils, monocytes, smooth muscle cells, and fibroblasts, whereas TGF- increases macrophage cytokine release and improves chemotaxis of fibroblasts and smooth muscle cells.

The very first 2–5 days of the leukocyte response are dominated by neutrophils, with macrophages taking control on the third day. Neutrophils have three major purposes. To begin, they create free radicals via the myeloperoxidase pathway to kill bacteria, which is a critical step for healing since wounds infected with bacteria do not heal correctly.

They also debride the wound by secreting proteolytic enzymes that break down non-viable tissue, such as serine proteases and matrix metalloproteinases (MMP-2 and -9). As a result, neutrophils phagocyte the dead bacteria as well as the residual matrix. When their activities are accomplished, they normally die and are cleaned by macrophages.

Monocytes start migrating to the wound and eventually develop into macrophages. In the inflammatory response, they become one of the major regulatory cells. M1 phenotype macrophages are related to the phagocytic activity of residual bacteria, sequestration, and synthesis of pro-inflammatory mediators in the early phases of inflammation.

After this time, M1 transforms into M2, demonstrating the reparative nature of macrophages. M2 macrophages are engaged in anti-inflammatory mediator production and tissue cleaning. These cells eliminate dead neutrophils, inactive host cells, damaged matrix, and foreign debris.

To enhance and resolve inflammation, M2 cells release cytokines, growth factors, and other mediators such as TGF-, TGF-, basic fibroblast growth factor (-FGF), PDGF, and vascular endothelial growth factor (VEGF).

Macrophages control the proliferative stage of healing by driving fibroblasts, keratinocytes, and endothelial cells to differentiate, proliferate, and migrate, resulting in new ECM deposition, re-epithelialization, and wound vascularization. The proliferative stage's purpose is to reduce the amount of constricted tissue and fibroplasia while also generating a functional epithelial barrier to activate keratinocytes. TGF-stimulated fibroblasts develop into myofibroblasts rich in alpha-smooth muscle actin and can multiply pseudopodia in the ECM, connecting fibronectin and collagen. This occurrence causes wound contraction, which aids in the healing process by allowing the wound's margins to approach.

Those cells additionally make ECM molecules (collagen, fibronectin, glycosaminoglycans, proteoglycans, and hyaluronic acid), which interact with the cells to regulate migration, growth, and differentiation. The wound closure, which includes angiogenesis, fibroplasia, and re-epithelialization, characterizes this stage.

The last stage of wound healing is characterized by the production of scar tissue and the growth of new epithelium, a process known as remodeling, which can take a year or longer. The remodeling is aimed at attaining maximal tensile strength by restructuring.

In ECM degradation and resynthesis, collagen strands expand in diameter in tandem with intracellular matrix development, whereas hyaluronic acid and fibronectin break down. The wound's traction force gradually increases with collagen deposition, and these fibers can regain around 80% of the force compared to normal tissue, but the original tissue's force can never be restored.

As established, several variables can influence cell connections and signaling processes throughout the wound healing process, either favorably or adversely. Plant-derived chemicals are among these variables, and they can aid the healing process via a variety of methods.

5.4. PLANT-DERIVED COMPOUNDS

Medicinal herbs have long been used as traditional therapy for a variety of disorders because they include phytochemicals, which are nonnutritive compounds found in plants that promote tissue regeneration and function as pro-angiogenic agents for wound healing.

Furthermore, bioactive plant compounds pique scientific and commercial interest in the synthesis of novel pharmaceuticals. Plants, on the other hand, provide a source of various macromolecules, including carbohydrates and proteins, which are widely employed as biomaterials in wound healing applications.

Figure 5.2. A woman is drying diverse medicinal plants and herbs in her courtyard before processing.

Source: Image by Flickr

5.4.1. Essential Oils

These natural oils (EOs) are perhaps the most abundant type of secondary metabolite, consisting of a complex of monoterpenes (10 carbons) and sesquiterpenes (15 carbons) that are primarily involved in plant defense mechanisms.

They are also known as volatile oils or aromatic plant essences and may be found in a variety of plant tissues such as flowers, leaves, barks, and so on. They have been widely used in cosmeceuticals and dermaceutical products after being obtained by aqueous extraction, steam distillation, or cold pressing in the case of citric fruits.

Figure 5.3. An aromatic plant.

Source: Image by Pixabay

The anti-inflammatory and antibacterial qualities of EOs make them particularly useful in the healing process. The effectiveness of essential oils in preventing bacterial proliferation, particularly antibiotic-resistant strains, yeasts, and filamentous fungi, has fueled research into their antibacterial action.

Some oils isolated from medicinal plants have shown therapeutic promise in fighting biofilms, a virulence mechanism formed by antibiotic-resistant pathogenic bacteria. Carvacrol and Thymol, for example, are monoterpenes found in essential oils of the Origanum species that have antibacterial and antifungal properties, as well as analgesic properties.

Several investigations have found that interactions amongst EO components, even at low concentrations, can result in antagonistic, additive, or synergistic effects. Although EOs are a mixture of plant molecules with different uses such as antibacterial and anti-inflammatory capabilities, in addition to possible therapeutic effects, as noted above, we focus on plant polysaccharides and lectins here.

5.5. CARBOHYDRATES

Basically, seeds play a crucial part in certain species' reproductive methods and constitute a critical stage in the life cycle of plants. They also play an important role in food and human health, which encourages science and technology to investigate a wide range of possible uses. These applications have made significant contributions to human well-being and health, including the invention of biopharmaceuticals.

Most cells in higher plants have a complicated system of polymers in their cell membrane, comprising cellulose, non-cellulosic polysaccharides (pectin), structural glycoproteins, and, on the secondary wall, lignin.

The presence of cellulose, which is composed of glucose chains connected to is a unique feature of plant cell walls. A range of hemicelluloses, which are polysaccharides with properties comparable to cellulose, are discovered crosslinked to such microfibrils.

5.5.1. Plant Cell Wall Polysaccharides

Cell membrane polysaccharides are classified as structural polysaccharides or storage polysaccharides. Primary and secondary walls both include cellulose and hemicelluloses, pectin, enzymes, and structural proteins, whereas secondary walls typically contain lignin but no proteins or pectin.

Secondary cell walls emerge when the cell's growth is interrupted and frequently display intricate specializations, with the inclusion of lignin being the most distinguishing feature. As a result, the secondary walls of cotyledonary and endospermic cells in many species' seeds lack lignin and contain little cellulose.

In contrast, the cell wall of storage tissues (endosperm or cotyledon) in certain seeds is relatively thick and includes deposits of polysaccharides that are mobilized following germination. Cell wall storage polysaccharides (CWSPs) are composed of mannans, galactomannans and glucomannans, and xyloglucans and Galatians.

Storage polysaccharides are primarily water soluble and create viscous, stable dispersions, absorbing a lot of this solvent. It keeps water around the embryo during imbibition and germination, protecting it from dehydration.

5.5.2. Galactomannans

Galactomannans are polysaccharides present in the neutral cell walls of dicotyledonous seeds. They serve as a storage mechanism, frequently being catabolized to deliver energy and carbon skeletons to the plant during germination.

They are more plentiful in seeds of the Leguminosae family, of which locust bean (*Ceratonia siliqua*), guar (*Cyamopsis tetragonoloba*), and tara (*Caesalpinia Spinosa Kuntze*), and fenugreek are the four largest commercial sources (*Trigonella foenum-graecum* L.).

Galactomannans are heterogeneous polysaccharides with a linear chain of D-mannopyranose residues linked by -glycoside linkages and D-galactopyranosyl joined by -type glycosidic bonds. Despite this structure, galactomannans are also known as hemicelluloses, and changes in the Gal/Man ratio produce substantial changes in the physicochemical properties of this natural polymer, such as average molecular weight, and intrinsic viscosity, and polydispersity. Furthermore, the solubility in water is substantially influenced by the sugar ratio, which varies depending on the source and separation process. The more the main backbone is replaced by galactosyl residues, the more soluble the galactomannan is in water.

5.5.3. Xyloglucans

Xyloglucans are polysaccharides present in the main cell wall of the cotyledon of many seeds that have both structural and storage functions.

Their primary chain is made up of D-glucopyranose connected by and branched in O-6 by -D-xylopyranoside units, which may also be replaced in O-2 by -D-galactopyranosyl units.

Figure 5.4.*Cotyledon tomentosa* in cultivation.

Source: Image by Wikimedia Commons

This class of polysaccharides is critical in regulating cell growth. Xyloglucans were shown to be connected with microfibrils in the experiments, suggesting that they, like other hemicelluloses, can give mechanical resistance and physical integrity to the intricate arrangements in plant cell walls.

In relation to this cellulose linkage, xyloglucans are joined via hydrogen bonds, and their long polysaccharide chains ensure the preservation of network microfibrils in cell wall growth. The common structure of storage xyloglucans, which allows them to form hydrogels and film solutions may be shaped into wound dressings capable of carrying potential healing molecules.

5.5.4. Exudate gums (Arabic, tragacanth and cashew gum)

Exudate gums are hydrocolloids with a high molecular weight and viscous appearance that are generated from the exudates of some plant species' branches and bark on their trunks. To get these molecules, a process known as gummosis must occur, which is triggered as a physiological defensive response to chemical, physical, and biological stimuli.

Chemically, they are composed of a complex structure similar to that of arabinogalactans, galacturonans, glucoronomannan, or glucoronomannan of acid nature, which is branched and replaced by major elements (C, H, O, and N), inorganic ions, and secondary metabolites synthesized via the phenylpropanoid pathway (tannins, terpenoids and other phenolic compounds).

Arabic gum (AG) is a polysaccharide containing complex and branching structures (adhesive and cohesive qualities) made of side and main chains with (1,3) and (1,6) and D-galactopyranosyl, -L-arabinofuranosyl, -L-rhamnopyranosyl, and -D-glucopyran units. GA is covalently coupled with protein fractions and a high amount of hydroxyproline, leucine, serine, and proline residues in some circumstances.

Figure 5.5. Pieces of raw Gum Arabic.

Source: Image by Wikimedia Commons

Among its known pharmacological capabilities are its abilities as a mucosal and intestinal anti-inflammatory, antibacterial, and antioxidant, biochemical components that are likely to influence the process of wound healing.

Tragacanth gum (TG) is an anionic acid branched heteropolymer with residual units of arabinose, glucose, xylose, galactose rhamnose, fucose, and galacturonic acid, and TG has important biological properties, such as biodegradability and biocompatibility, making it suitable for the design of hybrid scaffolds with pharmaceutical applications, the development of polymeric systems for controlled drug release, and guided tissue regeneration.

Cashew tree gum, like the polymer gums discussed above, is derived from the Anacardium genus species, which are plentiful in the Brazilian Northeast. The molecular structure consists of a -galactose (1–3) main chain and branched bonds (1–6), with side residues of glucuronic acid, 4-O-methyl glucuronic arabinose, rhamnose, xylose, glucose, and mannose.

The biochemical features connected with biological activities exhibit anti-inflammatory capabilities, oxidative stress and reactive oxygen species regulators, and antibacterial as well as gastroprotective effects.

5.6. PROTEINS

5.6.1. Latex Proteases

Certain species of plants have laticifer ducts that generate and store latex. This liquid contains a lot of natural chemicals such as secondary metabolites, glycosides, and proteases. Many researchers have investigated latex proteins as novel natural substances for biological purposes.

Cysteine and serine peptidases, often known as latex proteases or latex peptidases, are two of the most prevalent macromolecules derived from proteins. These macromolecules function in synergy with other latex sap proteins to offer the first line of defense against natural enemies in plants.

Proteases are generally found in both humans and animals, and their proteolysis functions, specificity, and bioactivity have rendered them well-known in the medical and pharmaceutical industry fields. Proteases such as metalloproteinases are endogenously secreted by fibroblasts, macrophages, mast cells, and endothelial cells after extracellular matrix damage in human biological systems.

These enzymes first contribute to the inflammatory phase of healing by debriding the wound necrotic tissue and subsequently in the cicatricial process by contributing to collagen remodeling and scar tensile strength lowering. Proteases and their inhibitors also contribute to ECM breakdown and deposition, resulting in a delicate balance required for appropriate and synchronized cutaneous wound repair.

Modulating ECM proteases using laticifer proteins has been utilized to improve the efficacy of healing processes in both acute and chronic wounds.

Recent advances in plant latex biotechnology have aided in the investigation of the pharmacological properties of protease-rich fractions of *Calotropis procera* latex, revealing its potential role in procoagulation and blood clot hydrolysis, modulation of inflammation, and enhanced wound healing in animal models using polyvinyl alcohol biomembranes as a vehicle for releasing laticifer proteins.

Furthermore, in experimental excisional wound models, a phytomodulatory galactomannan-based hydrogel was successfully employed to transport latex proteases from *C. procera.* A synergistic interaction was identified between galactomannan and proteases macromolecules, promoting repairing.

5.6.2. Lectins

Lectins are proteins or glycoproteins found in nonimmune nature that can uniquely detect and reversibly bind carbohydrate moieties without changing the covalent structure of their glycosyl ligands. This appealing feature separates lectins from other carbohydrate-binding proteins and enzymes.

They are also extensively spread in the plant world, often originating from leguminous seeds, and have critical roles and functions in biological processes such as molecular recognition, storage proteins, and plant defense mechanisms.

Their interactions with glycosyl ligands are mostly mediated by hydrogen bonds, van der Waals forces, hydrophobic contacts, and less commonly, electrostatic interactions. We focus here on lectins from jackfruit, breadfruit, and chempedak and their biological uses.

While lectins are widely dispersed in nature (animals, insects, viruses, fungi, and bacteria), the majority have been described from plant protein extracts, reflecting the simplicity of extraction and relatively high yields, often using a simple one-step affinity chromatographic approach.

Following the discovery of jacalin, new lectins with strong similarities to it were assigned to a family of jacalin-related lectins (JRL), which is currently separated into two distinct subgroups. The first group consists of galactose-specific lectins (gJRL) and a few additional Moraceae lectins that have galactose selectivity and are made up of subunits with a short chain and a long chain.

The second is the mannose-binding subgroup (mJRL), which is found in several plant families and consists of lectins with a unique affinity for glucose/mannose. Residues containing the binding subunits enclosed within a single polypeptide chain.

Lectins are naturally occurring bioactive proteins and glycoproteins with the potential to selectively bind carbohydrates. These non-immune sugar-binding proteins can agglutinate cells or precipitate glycoconjugates. They are abundant in nature and are known to perform critical roles in numerous biological processes.

Numerous lectins have previously been identified in plants, algae, fungus, invertebrates, and bodily fluids of lower vertebrates. Lectins may be utilized as models to investigate protein-carbohydrate interactions as well as a delicate tool for analyzing free form, lipid-bound, or protein-bound carbohydrates.

Because of their carbohydrate-binding selectivity, lectins are also utilized to transport medicines to the site of action. Many studies have demonstrated the unique properties of plant and animal lectins as recognition molecules in cell-molecule and cell-cell interactions in a variety of biological systems

Furthermore, they serve an important role in understanding biological processes, clinical diagnostic systems, and carbohydrate structure.

5.6.3. Plant lectins

Plant lectins have at least one non-catalytic domain that can bind to a particular mono- or oligosaccharide reversibly. Because of their widespread distribution and simplicity of separation, they were the first proteins to be examined.

500 distinct plant lectins have been identified and described to date. These lectins are typically found in seeds, roots, and leaves. They can aid in the recognition of glycoconjugates on the cell surface, as well as the separation and structural study of glycoproteins and oligosaccharides.

Furthermore, these lectins have been revealed to be extremely important in host-pathogen contact, development, cell signaling, and cell-cell communication. They also defend the plant against hazardous phytopathogenic microorganisms, insects, and predatory animals. Plant lectins are also essential for establishing symbiotic relationships with host plants and nitrogen-fixing bacteria.

Prior to the identification of Nod factors, lectins extracted from legume plant seeds were thought to operate as an intermediate between two symbiotic players. Legume lectins recognize and bind to carbohydrate moieties on the bacterial surface. This interaction results in either agglutination of bacteria far from the root or adhesion of bacteria to root epithelial cells. Bacterial contact with root hairs increases the formation of infection threads, which are required for the growth of an efficient root nodule.

For example, when the PSL (*Pisum sativum* lectin) gene was introduced into *Trifolium repens* (white clovers) via *A. rhizogenes* transformation, active nodules were generated. Further research has focused on the effect of cross-species lectins in promoting nodulation via their rhizobia in the host plant. For example, inserting the SBL (Soybean lectin) transgene into *Lotus corniculatus* (usually modulated by *Mesorhizobium loti*) improved its binding affinity with *Bradyrhizobium japonicum*, the rhizobia linked to soybean.

Figure 5.6. Leaves of *Pisum sativum*.

Source: Image by Wikimedia Commons

Similarly, insertion of PSL transgene or *Glycine max* lectin apyrase/ GSC2 into the root of transgenic rice (which generally establishes a symbiotic association with mycorrhiza) resulted in root colonization by different rhizobia (*R. leguminosarum*, *B. japonicum*, and *Rhizobium* species NGR234) as compared to control roots.

It has also been shown that introducing the GSC2 gene into *L. japonicum* increases nodule development and infection thread advancement after inoculation with *Mesorhizobium loti*. In addition to the aforementioned, LecRK DB46, also known as LNP (lectin nucleotide phosphohydrolase) has been demonstrated to impact nodule formation, and its expression level rises during nitrogen-limiting circumstances.

5.6.4. Artocarpus lectins

Artocarpus is a genus of roughly 60 trees and shrubs of Southeast Asian and Pacific origin that belong to the Moraceae family; all species are lactiferous, with milky sap produced by the leaves, twigs, and stems. The name is a combination of the Greek words artos (bread) and karpos (fruit).

Although most Artocarpus species, such as *A. hypargyreus* (kwai muk), *A. lakoocha* (lakoocha), *A. kemando* (pudau), *A. hirsutus* (anjily), *A. chama* (chaplaish), and *A. odoratissimus* (marang), are restricted to Southeast Asia, several species are widely distributed and cultivated throughout the tropics due to their edible These include *A. heterophyllus* (jackfruit), *A. altilis*, and *A. integer* (cempedak, also known as chempedak), all well-known species that provide as significant sources of plant lectins that may be easily restored from seed flour.

5.6.5. Bacterial lectins

These also are known as adhesins because they aid in the adhesion of bacteria to host cells during infection. Through the carbohydrate-recognition domain (CRD), they bind to glycan receptors. Most bacteria have numerous adhesins with different carbohydrate specificities.

Some adhesive signs link to terminal sugar residues through CRD, whilst others bind to internal sequences of linear or branching oligosaccharide chains. It is crucial in determining the tropism of the symbiont or pathogen during contact with host glycans. These lectins aid in adhesion and symbiotic relationships.

5.6.6. Fungal lectins

Mucins and N-acetyl galactosamine (GalNAc) residues are very unique. Different fungal lectins have been found, with mushrooms accounting for 82%, microfungi (molds) accounting for 15%, and yeasts accounting for 3%. percent With a few exceptions in mycelia, they are mainly found in fruiting bodies. They are required for mycorrhization growth, development, morphogenesis, and molecular recognition. They also play a role in early infection by interacting with host glycoconjugate.

5.6.7. Jackfruit (jacalin, ArtinM and jackin)

In traditional medicine, the species has a long history of usage, including treatment of asthma, dermatitis, anemia, diarrhea, and fever; antisyphilitic and anthelmintic characteristics; calming effects in convulsions; and wound healing capabilities.

Because of extensive research on the phytochemical and pharmacological qualities of all sections of the plant, jackfruit is in great demand in sectors such as cosmeceutical, pharmaceutical, and natural food processing for supplement markets (pulp, leaf, root and bark).

Figure 5.7. Jackfruit Tree.

Source: Image by Flickr

Jackfruit extracts and metabolites include various important bioactive components that may have multiple health-promoting benefits for heart and skin problems, as well as ulcers and cancer.

Furthermore, current research on the qualities of jackfruit has shown further biological discoveries associated with antibacterial, antitubercular, antiviral, antifungal, antiplatelet, and antiarthritic effects, indicating therapeutic possibilities.

Jacalin, a D-galactose-binding lectin from *A. integrifolia* seeds, was discovered to account for more than half of the protein in jackfruit crude seed extracts in 1979. This is also true for galactose-binding lectins like frutalin and CGB from the Artocarpus genus (chempedak galactose-binding).

Jacalin is a 65 kDa tetrameric two-chain lectin with a heavy chain of 133 amino acids and a light chain of 20–21 amino acids that combine to produce a 3D shape as a single globular unit.

This lectin displays two bands on SDS-PAGE between 20 and 14 kDa, corresponding to glycosylated and mildly or non-glycosylated forms, respectively. Gly1, Tyr78, Val80, Gly121, Tyr122, Trp123, and Asp125 are the key residues involved in jacalin's carbohydrate-binding site (CBS).

The O4 hydroxyl group of the galactose axial position establishes hydrogen bonds with the side chain of Asp125 and the terminal amino group of Gly1 in D-galactose-jacalin complexes. Asp125 can still interact with O4 at the equatorial position, as in glucose and mannose, but not with the amino group.

This explains jacalin's high selectivity for galactose over glucose and mannose at the major binding site. Furthermore, the elimination of the "T-S-S-N" peptide linker and post-translational cleavage provide a greater hydrogen bond joining the - and -chains, since non-cleavage leaves a neutral peptide NH group.

Following the discovery of jacalin, additional research discovered that jackfruit seed extracts contain trace levels of a D-mannose-binding lectin.

The word artocarpin was temporarily substituted by KM+ because the lectin had gotten confused with various compounds from Artocarpus spp. that were similarly classified. Furthermore, the galactose-binding lectin in *Artocarpus lakoocha* seeds was named artocarpin.

The designation KM+ is derived from the various affinity chromatography processes used to extract jacalin from immobilized D-galactose matrices (retained fraction J). K denotes the unretained fraction, whereas M+ is the retained fraction on immobilized mannose matrices.

Nonetheless, there was some uncertainty about the accepted nomenclature, which prompted the recommendation for a reasonable name change to ArtinM based on the universal code for plant proteins. Following the discovery of jacalin, more research indicated that the jackfruit seed takes into consideration both the origin of a lectin and the specificity of sugar recognition.

Following that, we will use ArtinM to add early work and KM+ mentions (artocarpin). ArtinM is a single polypeptide of 150 amino acids with four isolectins and a pI range of 5–6.5 that has 52 percent sequence similarity with jacalin. Unlike jacalin, there are no aromatic residues on ArtinM's

CBS, which consists of Gly15, Asp138, Leu139, and Asp141. Indeed, it is thought that jacalin's galactose specificity was obtained by a two-step process using ArtinM as a possible precursor: mutation of crucial aliphatic residues near the sugar-binding pocket to aromatic ones, and then breakage of a short loop, which forms a positively charged N-terminal that interacts particularly with O4 in the axial location.

Because of its affinity for chitin, jackfruit seeds contain a third lectin called jackin, in addition to jacalin and ArtinM. Nonetheless, its limited yield from natural sources continues to impede further characterization, but this may soon be solved by high-yield heterologous manufacture in microbial systems.

5.6.8. Breadfruit

The Pacific Islands are indeed the birthplace and diversification center of breadfruit (*A. altilis*), sometimes known as *A. communis* or *A. incisa*. The species evolved from a seeded, diploid parent, *A. camansi*, yielding two varieties:

One-seeded (var. seminifera) with pronounced rind spines similar to jackfruit, and one seedless (var. apyrena) with a spineless outer layer. When cooked, the latter is well-liked by native Brazilians because of its high starchy content of compound fruits with high quantities of minerals and provitamin A carotenoids. Breadfruit flour was recognized as a food Generally Recognized as Safe (GRAS) by the US Food and Drug Administration (FDA) in 2016. *A. incisa* seeds discovered lectins that behaved similarly to those identified in jackfruit seeds in 1983.

Figure 5.8. Breadfruit tree.

Source: Image by Wikimedia Commons

Breadfruit seeds have a high-water content (up to 60%) and a modest protein level (12.25 percent of dry weight). The majority of this protein is obtained as frutalin by chromatographic techniques utilizing crude seed flour extracts.

As a result, frutalin is the most common lectin in this species, with multiple-binding capabilities that allow the same CBS to identify a variety of different ligands, although having greater affinities for -D-galactose monosaccharides and complex carbohydrates including Gal1–3 glycans.

Hydrogen bonding through O1, O3, O5, and O6 and backbone/side chain hydroxyl groups dominate the CBS of frutalin in galactose binding. The CBS of frutalin, like the Moraceae lectins, is situated in a domain towards the N-terminus of the chain and, which consists of four important residues.

Gly25, Tyr146, Trp147, and Asp149 are all amino acids. Around ten interactions occur, including the C1 hydroxyl to Tyr146, the C3 hydroxyl to Gly25, the C4 hydroxyl to Gly25 and Asp149, and the C6 hydroxyl to Tyr146, Trp147, and Asp149.

Furthermore, there is evidence that frutalin has stereospecificity, capable of binding -D-galactose specifically, because it was previously isolated on a cross-linked galactomannan column but not on -galactosyl-immobilized matrix. The second most prevalent lectin in breadfruit seed extracts is rutapin (FTP).

FTP research began in 1998, with the identification of three separate lectins with unique carbohydrate recognition within the same species. Nevertheless, additional research proved challenging since native FTP was limited by poor yields and contamination with Frutalin, a significant problem because frutalin binds a variety of sugar moieties.

Does have a high concentration of plant extracts Recombinant FTP synthesis is now a viable solution to this dilemma, allowing large-scale heterologous expression to provide continuous supplies for further characterization and potency improvement, especially in biomedical applications.

FTP is a hololectin, described as a homotetramer with an identical CBS per protomer, capable of binding either similar or slightly structurally related sugars. The CBS is composed of the following residues:

Gly16, Asp139, Leu140, and Asp142, which are dispersed in another few loops linking the strands 5 and 6, 7 and 8, and 11 and 12. Several hydrogen bonds (HB) form in FTP-glucose and FTP-mannose complexes involving the carbohydrates Gly16, Leu90, Gly138, Asp139, Leu140, Asp142, and O3, O4, O5, O6.

In MD simulations, Lys60 plays a crucial role in creating salt bridges with Asp139 in FTP-glucose complexes, lowering the interaction between this former residue and mannose and limiting the repulsion of the mannose hydroxyl groups with oxygen. Mannose was totally bound in this scenario. Mannose was more entirely surrounded in the carbohydrate-binding site and was additionally stabilized by indirect contact with Asp139 via water molecules.

This local structure is more persistent in the case of mannose than glucose, indicating that FTP has a greater affinity for mannose residues than glucose. Subsequent research on breadfruit seeds identified frutackin, a lectin related to jackin.

Both lectins have a 14 kDa polypeptide chain composed of three chains connected by disulfide bonds, and their partial amino acid sequences exhibit similarities in terms of molecular mass, secondary structure, and primary sequence, but not to other plant chitin-binding proteins. *F. moniliforme* and *S. cerevisiae* are both inhibited by jackin and frutackin.

5.6.9. Chempedak

Though sometimes known as chempedak or chempedak in the tropics, *Artocarpus integer* (Thumb.) Merr. is native to India and has fruits similar to jackfruit. The *A. integer* species is high in phenolic compounds and exhibits potent cytotoxicity against murine leukemia P-388/.

Furthermore, when applied to wounds, chempedak paste of the inner bark inhibits infection and promotes healing, as does heated leaf extracts. *A. integer* seed flour extracts include significant quantities of Chempedak galactose-binding lectin (CGB).

The bioactivity of the lectin was discovered when extracts were evaluated for selective stimulation of peripheral blood mononuclear cells; at 20 g/mL CGB induced T-lymphocyte proliferation. CGB, like frutalin and jacalin, is translated as a propeptide and subsequently post-translationally processed into a normal gJRL lectin with a 13-kDa (133 amino acid) -chain and a 2.1-kDa -chain (21 amino acids).

CGB differs from jacalin by six amino acids, resulting in 97 percent similarity in their amino acid sequences. The contacts are largely preserved, displaying the same CBS as in jacalin complexes.

The O atoms on the sugar ring are coupled with the sidechain and main-chain N and O atoms on the chain in Gal-CGB complexes (O3 and Gly1 N; O4 and Gly1 N and Asp125 OD1; O6 and Trp123 O, Trp123 N and Tyr122 N; and O5 and Tyr122 N).

Figure 5.9. Chempedak (chlebowiec chempedak *Artocarpus integer*).

Source: Image by Wikimedia Commons

Likewise, GalNac-CGB complexes include bound disaccharides in the same area via hydrogen bonds (O3 and Gly1 N; O4 and Gly1 N and Asp125 OD1; and O6 and Asp125 OD1). Tyr78, Gly121, and Tyr122 also contribute to a variety of hydrophobic interactions.

Despite being structurally similar to jacalin, CGB was unable to bind mannose, as determined by isothermal calorimetry and co-crystallization investigations. This shift in CGB specificity is thought to be produced by small alterations in the environment surrounding the sugar-binding site, such as solvent molecules.

CMB, also known as chempedak lectin-M, is found in extracts of *Artocarpus integer* seeds. In crude extracts of *A. heterophyllus* seeds, the lectin is 20-fold more prevalent than ArtinM. CMB is a 64-kDa tetramer, with certain polypeptides disulfide-linked to form dimers, according to structural research.

CMB's functional activity was evaluated by analyzing interactions with distinct human immunoglobulin isotypes: high binding to IgE and IgM was observed, in contrast to CGB and jacalin, which strongly interact with IgA1. The lectin has a similar carbohydrate binding selectivity as ArtinM, with

Man—1–3Man being the most powerful inhibitor, followed by methyl—D-mannopyranose and D-mannose.

The Artocarpus genus appears to use a variety of lectins, albeit few of these lectins have been identified yet. have been processed and functionally assessed Nonetheless, Artocarpus seeds consistently contain more than one lectin with different carbohydrate recognition properties. Overall, the JRL family is complicated, having a wide range of biochemical characteristics and activities that have sparked widespread interest due to its critical medicinal applications.

Plant lectins are intriguing candidates for selectively modulating immunological responses in plants due to carbohydrate-binding interactions with cell wall glycoproteins.

As a result, it is critical to understand the chemical intricacies of lectin binding to CBS domains and how cellular immunological signaling is activated downstream. In this respect, it is worth noting that, despite their great sequence and structural similarities, Artocarpus lectins exhibit a diverse array of biological functions.

5.7. RECENT ADVANCES USING PLANT BIOMATERIALS FOR WOUND HEALING

The growing number of chronic-wound patients globally has fueled an aggressive push in the wound-care business to find low-cost, effective wound healing technology. As a result, most of this knowledge is patentable, which plays a significant role in recognizing technological growth patterns.

Major efforts made in the previous five years have aimed to create biomaterials that use plant macromolecules as a source of biomolecules with promise for wound healing applications.

5.7.1. Nanomaterials for Application in Wound Healing

Nanomaterial-based wound healing is an important tool for treating and preventing wound infections, with several advantages over standard-of-care treatments (SOC). Wounds are a "hidden pandemic" that reduces patients' quality of life (QoL). An economic review of acute and chronic wounds in 2018 revealed that almost eight million beneficiaries had at least one form of wound or associated infection. Exudate-absorbents with high swelling capacity and porosity, a high-water vapor transfer rate (WVTR), antimicrobial, and anti-inflammatory characteristics are ideal wound dressings. They offer good elasticity and flexibility, drug loading capacity, tensile strength, and spreadability, and produce a moist wound environment that speeds healing, however, most commercial dressings lack some of these properties.

It includes the current dressings, along with their benefits and drawbacks in wound healing. Traditional wound treatment (e.g., herbal medications, honey, bandages, and dressings) left scars regardless of cosmetic or functional adjustments. In wound healing, honey has intriguing immunostimulatory, antibacterial, antioxidant, and anti-inflammatory properties.

Nevertheless, honey has various drawbacks or negative effects when used topically in wounds. Honey-based dressings, for instance, could be difficult to prepare; extreme temperatures start making it even more fluid; a transient stinging feeling might indeed occur; this could boost blood glucose concentration in diabetic patients in large wound areas; the excessive application may dehydrate tissues, and pollen/bee proteins in honey may cause hypersensitivity.

Other than restitution ad integrum (the delay in restoring tissue integrity), the constraints of these conventional materials worsen wounds, particularly chronic wounds. As a result, possible wound healing materials might improve clinical results.

Innovative polymeric nanofibers, polymeric nano scaffolds, and nanoceria have evolved for wound healing management. Due to their high surface area-to-volume ratios and nano size, natural origin nanomaterials and drug delivery vehicles are well suited for cellular responses, penetrability, and active drug delivery in wound healing.

Silver nanoparticles (AgNPs) functionalized bio cellulose acts as an efficient covering against Gram-negative bacteria and speeds up open wound healing. Silver nanoparticle-coated polyester-nylon dressings were extremely biocompatible, had antibacterial activity, aided normal human cell growth *in vitro*, and showed normal biodistribution with little toxicity *in vivo*.

The existing evaluations may have solely focused on one or two nanotechnology systems in wound healing (e.g., diabetic ulcers). The various systems and use of nanoparticles in wound healing (i.e., inorganic nanomaterials, organic and hybrid nanomaterials, and nanofibers). We also discussed the key issues about the potential future usage of nanomaterials in wound healing.

5.7.2. Inorganic/organic nanocomposites in wound healing

Inorganic/organic nanocomposite scaffolds have piqued the interest of researchers because of their unique antibacterial and mechanical capabilities when an inorganic nanoparticle is combined with a supportive polymer matrix.

Because of their nature, their inorganic/organic material ratio, and the size and distribution of inorganic nanoparticles in polymer matrices, inorganic/organic nanocomposites have fulfilled the rising need for wound healing.

The *in vivo* activity of intermediate-modified gold nanoparticles (AuNPs) coupled with polycaprolactone (PCL)/gelatin nanofibers against multidrug-resistant (MDR) bacteria has been demonstrated. Copper sulfide (Cu2S) nanoparticle scaffolds electrospun with polylactic acid (PLA)/PCL polymers cure diabetic full-thickness skin lesions and dramatically induce angiogenesis *in vivo*.

Skin tissue engineering scaffolds are made from natural (e.g., dextran, chitosan, and alginates) and synthetic polymers (e.g., poly[-caprolactone] and poly [acrylic acid] [PAA]). Whereas natural polymers are biocompatible and enzymatically biodegradable, their strength is insufficient, and their degradability is unregulated.

Synthetic polymers have tunable structural, mechanical, and chemical characteristics. However, their utility in wound treatment is restricted due to their low biocompatibility, limited capacity to promote wound healing, loss of mechanical qualities, and the formation of hazardous chemicals during deterioration.

5.8. CONCLUSION

Methods of separating and identifying plant macromolecules differ in a number of ways from those used with low molecular weight constituents. Skin is the largest organ, which presents a fairly robust arrangement, working as a natural shield against physical, chemical, and bacterial damage to the body.

In a simple case, macromolecular constituents may be dissolved by homogenizing plant tissue with salt solution and then precipitated by changing the pH of the extract. There are also many detailed texts devoted to the methodology of isolating proteins and nucleic acids, mainly from animals but also from plant tissue. We focused our attention on plant macromolecules such as carbohydrates and proteins (in particular hemicelluloses and lectins) as biomolecules for wound healing applications. This section will be concerned mainly with the problems particularly associated with the isolation of macromolecules from plants, as opposed to, animal tissues.

Chemically, macromolecules consist of long chains of small structural units or 'building blocks, linked covalently in a number of different ways. Except for energy-related compounds, carbohydrates can be roughly divided into two categories: Carbohydrates as matter and carbohydrates as information.

The three main classes of macromolecules found in plants are thus proteins, polysaccharides and nucleic acids. This complex mechanism is dependent on many cell types and mediators interacting to maintain the physiological regulation of the skin.

Polysaccharides are similarly derived from the union of simple sugar units, such as glucose, joined through ether (-0-) links. Proteins, for example, are long chains of amino acids (up to twenty different ones) joined together through peptide (-CO-NH-) links. The role of carbohydrates can be demonstrated not only as targeting reagents but also as immune antigens and adjuvants. However, mixed polymers are also known, such as glycoproteins, which contain both sugars and amino acids in covalent linkage.

REFERENCES

1. Chen, H., Jia, Y. and Guo, Q., 2020. Polysaccharides and polysaccharide complexes as potential sources of antidiabetic compounds: A review. *Bioactive Natural Products*, [online] pp.199-220. Available at: <https://www.sciencedirect.com/science/article/abs/pii/B9780128194836000060> [Accessed 29 June 2022].
2. Domingos de Sousa, F., Rogênio da Silva Mendes, F., Jovanny Bermudez-Sierra, J., Fernanda Brandão da Silva, A., da Silveira Vasconcelos, M., de Fátima Goebel de Souza, T., de Oliveira Nunes, M., Eufrásio Vieira-Neto, A., Roberto Lourenzoni, M., Diógenes de Oliveira-Filho, R., Rolim Campos, A., de Azevedo Moreira, R. and Cristina de Oliveira Monteiro-Moreira, A., 2020. Plant Macromolecules as Biomaterials for Wound Healing. *Wound Healing*, [online] Available at: <https://www.intechopen.com/chapters/69102> [Accessed 29 June 2022].
3. Manivasagan, P. and Kim, S., 2014. Extracellular Polysaccharides Produced by Marine Bacteria. *Advances in Food and Nutrition Research*, [online] pp.79-94. Available at: <https://www.sciencedirect.com/science/article/abs/pii/B9780128002698000051> [Accessed 29 June 2022].
4. Mishra, A., Behura, A., Mawatwal, S., Kumar, A., Naik, L., Mohanty, S., Manna, D., Dokania, P., Mishra, A., Patra, S. and Dhiman, R., 2019. Structure-function and application of plant lectins in disease biology and immunity. *Food and Chemical Toxicology*, [online] 134, p.110827. Available at: <https://www.ncbi.nlm.nih.gov/pmc/articles/PMC7115788/> [Accessed 29 June 2022].
5. Su, L., Feng, Y., Wei, K., Xu, X., Liu, R. and Chen, G., 2021. Carbohydrate-Based Macromolecular Biomaterials. *Chemical Reviews*, [online] 121(18), pp.10950-11029. Available at: <https://pubmed.ncbi.nlm.nih.gov/34338501/> [Accessed 29 June 2022].

CHAPTER 6

Functional Applications of Macromolecules

CONTENTS

Polymers now are found in almost every aspect of our daily lives. Ever since the introduction of the first entirely synthetic plastic by Leo Baekeland in 1907, this field of study has expanded dramatically. New polymers were developed, and their properties were constantly enhanced.

Simultaneously, polymerization methods were created and improved in order to obtain control over, for example, the average molecular weight (Mn), dispersity (Đ), architecture, or usefulness of the produced polymers. However, the twenty-first century has brought new problems and new fields of research, such as sequence-controlled polymers.

6.1. INTRODUCTION

This phrase refers to polymers in which the monomer sequence is managed and the dispersity is decreased. However, control over the sequencing or the length of the polymer is not always total, and the characteristics of the resulting polymer might be influenced by the intrinsic molecular weight distribution.

Polymer chemists are now striving for more precise chemical methods in order to produce synthetic uniform macromolecules with unique features, inspired by sophisticated technologies found in nature, particularly the functioning of biomacromolecules such as DNA or proteins. As a result, a new discipline known as "sequence-defined polymers" has emerged.

The intense pursuit of ultraprecise, discrete (i.e., strictly monodisperse) synthetic macromolecules resulted in the invention of novel synthetic techniques or the reuse/adaption of previously recognized chemical instruments for their synthesis. Solid phase synthesis is one of the most extensively used processes.

Although this technology was originally designed for peptide synthesis, it has now been widely used to generate distinct synthetic oligomers. As a result, numerous chemical processes that were previously done in solution have been adapted to solid phase synthesis such as copper-catalyzed azide-alkyne cycloadditions, Passerini three-component reactions, and thiolactone chemistry.

The simplicity of purifying drives the interest in solid phase synthesis, but it has certain downsides as well. For example, chain length can be limited due to incomplete coupling stages or oligomer aggregation, restricting the accessibility of chain ends, and so on. As a result, several research groups choose solution phase synthesis techniques. The reaction can be carried out with the aid of soluble polymer support or directly in solution. Furthermore, several approaches in the solution have the advantage of being up-scalable.

6.2. CHAIN LENGTH LIMITATION AND UP-SCALABILITY ASPECTS OF DISCRETE SYNTHETIC MACROMOLECULES

The fields where discrete synthetic macromolecules are used in recent years, although it should be noted that each specific application has various requirements in terms of scalability and chain length of the individual macromolecules.

This means that while developing a unique synthetic method to produce tailor-made homogenous macromolecules, the nature of the desired application should always be examined. Certain applications demand a huge volume of product and/or lengthy macromolecular chains, whilst others require merely milligrams of oligomeric structures.

As a result, a quick review of several procedures that have the potential to be scaled-up or result in lengthy polymer chains is provided, while the reader is directed elsewhere for a detailed overview of the various synthetic processes. A popular strategy focuses on the iterative coupling of building pieces and can be accomplished directly in solution or with the help of a (soluble) solid support.

While the latter has many advantages for the development of specific macromolecules (for example, purification is limited to filtration, an excess of reagents can be used to increase conversion as well as reaction kinetics, and possible automation), milligram scale reactions in a research lab environment limit the applications to specific areas of research.

Despite their modest number, certain research investigations have shown that larger scales or longer macromolecules may be achieved efficiently using support-based methodologies.

Fuzeon, a peptide composed of amino acids, is manufactured on a multi-ton scale each year utilizing a combination of solid- and solution-phase techniques. Despite having a biological backbone, this macromolecule indicates that scaling concerns can be addressed in theory. Iterative procedures in solution, on the other hand, have less restrictions on the reaction scale but frequently need the use of labor-intensive purification processes such as column chromatography.

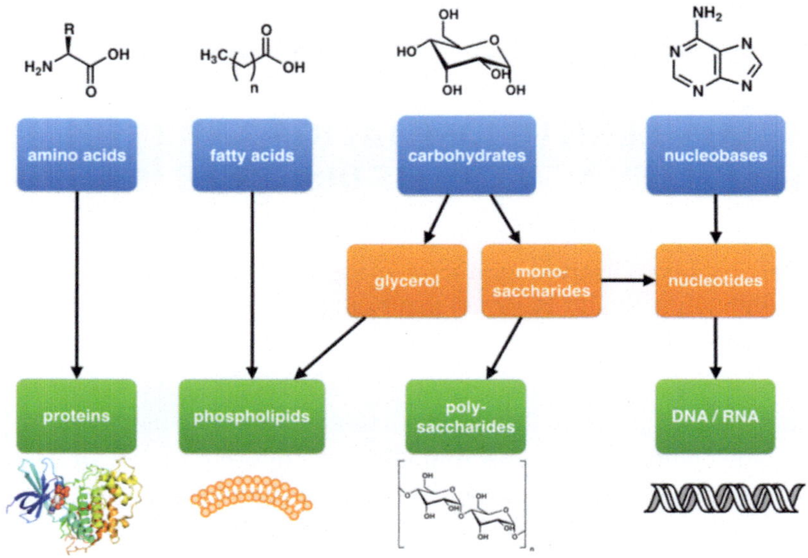

Figure 6.1. Basic molecules of life.

Source : Image by Wikimedia commons

Surprisingly, Meier and colleagues discovered that increasing the chain length facilitated the purification of distinct synthetic macromolecules using column chromatography for their specific technique. This seemingly contradictory discovery comes from the fact that an overabundance of low molecular weight chemicals may be easily eluted using a polar solvent, followed by the collection of the desired oligomer.

Gao and colleagues employed a polar-inversed technique to preferentially precipitate their uniform, positively charged macromolecules in the presence of unreacted neutral monomers, avoiding lengthy work-up processes. In another paper, Livingston and colleagues coupled a liquid phase method with selective molecular sieving to create homogeneous poly-ethers using simple sieving and extraction stages.

Porel and Alabi showed that a careful choice of building blocks might restrict the purifying procedure to a simple aqueous extraction. This method resulted in the multigram synthesis of monodisperse macromolecules, which were then utilized to create polymer networks through thiolene chemistry. To the best of the available knowledge, the latter paper is the first to use sequence-defined oligomers in a bulk material.

6.3. APPLICATIONS OF DISCRETE SYNTHETIC MACROMOLECULES IN MATERIAL SCIENCE

6.3.1. Macromolecular Data Storage

This has lately been said that the supply of microchip-grade silicon will not be able to keep up with the rising demand for storing all of the data that is produced every day, thus researchers are looking into alternatives to present data storage medium.

It is scarcely surprising that DNA is one of the alternatives, as nature has improved the process of storing a full genome in these biopolymers over the previous billion years, and DNA sequencing has been a topic of study for decades.

However, when employing sequence-defined synthetic macromolecules to encode information, some of the inherent limitations associated with DNA (e.g., hydrolysis of phosphodiester bonds, low Shannon capacity, costly and limited number of building blocks) may be addressed when employing sequence-defined synthetic macromolecules to encode information at the molecular level.

Furthermore, whereas DNA sequences are made up of four distinct nucleobases, the number of possible repeating units that may be included into sequence-defined macromolecules is far greater, possibly permitting the development of vastly denser data storage medium.

This may be shown as follows: Using a binary alphabet allows for the synthesis of 64 different hexamers, whereas ten different building blocks can already result in 106 different hexamers.

Boukis and Meier used this method to generate a "molecular alphabet" of 116 distinct building blocks by merging two multicomponent processes. This accomplishment eventually allowed them to store up to 24 bits of data per repeating unit. In a subsequent paper, the same researchers devised a synthetic technique based on the Passerini three component reactions in which the backbone and side chain of each individual repeating unit can be altered Due to the greater number of potential permutations, this dual sequence definition resulted in larger storage capacity when compared to more traditional techniques.

Similarly, Du Prez and colleagues used a similar method to encode a QR code onto a series of short sequence-defined oligomers. In this case, the use of fifteen different acrylates in their automated amine-thio-lactone-ene protocol avoided the need for longer macromolecular chains, which can be a difficult challenge when using a solid phase strategy to store a large amount of information because the backbone and side chain of each individual repeating unit could be varied.

6.4. SELF-ASSEMBLY OF DISCRETE SYNTHETIC MACROMOLECULES

6.4.1. Self-Assembly of Discrete Block Copolymers

Nature is full of interactions between molecules that result in higher-ordered structures. The assembly of phospholipids to create the cellular membrane or the construction of multiple copies of coat proteins in the outer shell of viruses, which gives them their characteristic forms, are two common examples.

The molecular assembly of tiny amphiphilic compounds has been intensively researched, and theoretical models based on molecular geometry and supramolecular interactions have been devised to interpret and predict their behavior. Block copolymers may also have a proclivity to assemble into well-defined morphologies.

This phenomenon is caused by incompatibility between the distinct blocks, which segregate to minimize interfacial energy. Similarly, theoretical and computational models have been established to describe this.

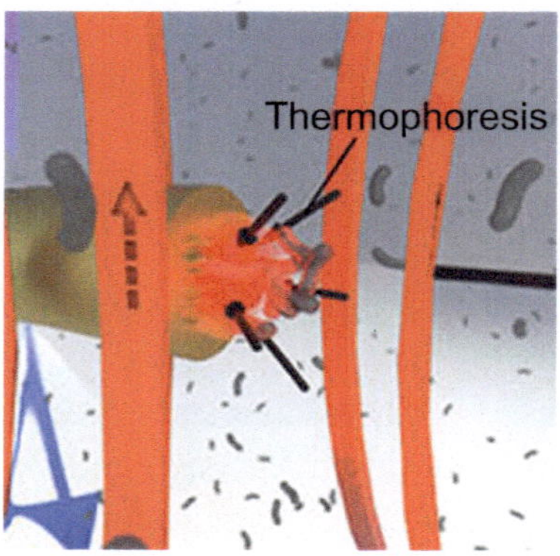

Figure 6.2. Plasmonic optical fiber for bacteria manipulation.

Source: Image by opg.optica.org

However, these models frequently presume discrete block copolymers, in contrast to the scattered nature of the real polymers employed in experimental research, which are generated using various live and controlled polymerization processes.

As a result, these models do not capture the entire image and do not account for variations in the chain length. Nonetheless, the continual growth of the area of sequence-controlled polymers has already permitted the production of discrete equivalents to conventional scattered commercial polymers.

Recent studies by the Meijer and Hawker groups have shown that even extremely modest dispersity can have a substantial effect on the self-assembly capabilities of block copolymers in the low molecular weight domain.

6.5. FOLDAMERS BASED ON UNIFORM MACROMOLECULES

The folding of biomacromolecules into secondary, tertiary, or even quaternary structures allows them to execute a wide range of sophisticated yet amazing activities like molecular recognition, transportation, information storage, and catalysis while utilizing a restricted repertory of building blocks (i.e., 20 amino acids and 4 nucleobases).

Furthermore, the folding of macromolecules into native states is a rather complex process that involves a variety of consequences.

On the one hand, favorable enthalpic interactions including intramolecular hydrogen bonding, electrostatic-, hydrophobic-, and van der Waals contacts encourage folding.

Building a secondary structure, additionally, is connected to restricting the level of freedom and is hence undesirable in terms of entropy. Therefore, folding could be viewed as a balance between antagonistic interactions, despite being a very cooperative process for most circumstances.

Most research has shown that the fundamental principles guiding the folding of proteins may be generalized to their synthetic equivalents throughout time. Indeed, other research groups began to investigate if uniform macromolecules based on abiotic building blocks were also capable of adopting special 3D structures that may enhance the distinctive qualities of biomacromolecules.

6.6. APPLICATIONS OF DISCRETE SYNTHETIC MACROMOLECULES IN LIFE SCIENCE

Nature uses controlled amino acid sequences to generate higher order structures such as -helices or -sheets that form peptides or proteins. Because the features of these biopolymers are entirely dependent on the correct sequence of building blocks, a focus on sequence control has proved harmful in biologically important applications.

Regardless of the fact that current research is mostly focused on the usage of narrow dispersity polymers derived from physiologically relevant monomers, examples from recent literature demonstrate many applications of discrete macromolecules in life sciences.

6.6.1. Antibacterial Properties of Discrete Synthetic Macromolecules

Though medicines have significantly enhanced life span throughout the last century, the emergence of antibiotic-resistant bacteria is among the most serious threats to human health. As a result, researchers have been seeking alternatives to standard antibiotics.

Antimicrobial peptides, which are expressed by the immune systems of all multicellular animals, have motivated their efforts. All of those are short peptides, with 88 percent of the 3180 known structures including fewer than 50 amino acids. While these peptides have a wide range of structural characteristics, many of them have comparable physicochemical features.

They have a net positive charge due to cationic amino acids (e.g., lysine or arginine), but they also have a substantial proportion of hydrophobic amino acid residues, which are essential for their action.

Due to their positive ions, they may stick and deposit semi-selectively on a bacterial cell membrane, which is richer in negatively charged phospholipids than mammalian cells.

Therefore, the insertion of hydrophobic moieties has the ability to rupture the membrane, destroying the bacterial cells. The use of antimicrobial peptides as antibiotics is hampered by their high manufacturing costs and susceptibility to proteolysis. As a result, synthetic mimics that mirror these critical physicochemical features have been produced.

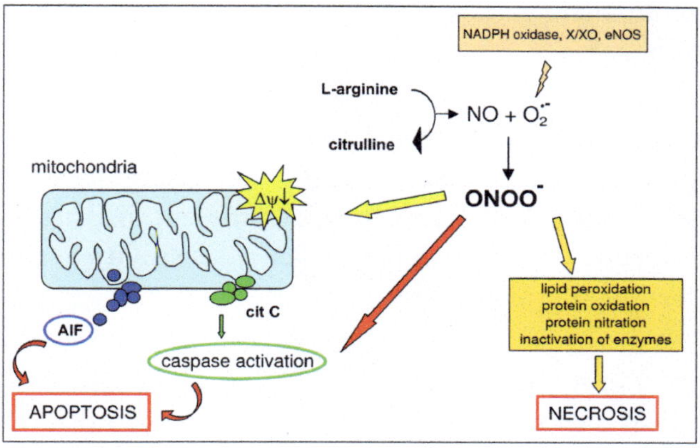

Figure 6.3. Reactions of per-oxy-nitrite leading to either apoptotic or necrotic cell death.

Source: Image by Wikimedia commons

The antibacterial efficacy of these mimics is frequently tested by measuring the minimum inhibitory concentration (MIC) required to suppress bacterial multiplication. However, it is also necessary to analyze the toxicity on eukaryotic cells.

6.7. OTHER APPLICATIONS OF DISCRETE SYNTHETIC MACROMOLECULES

Meier and colleagues, in particular, used repeated synthesis to create conjugated, rod-like pentamers from oligo (phenylene ethynylene). Although the photophysical characteristics of the resultant macromolecules differed only slightly, the differences in hydrodynamic volume and thermal properties were more pronounced.

The research might be expanded to include longer sequences beyond a pentamer, which could have an impact on the photophysical characteristics. Despite all the positive outcomes, the macromolecules were synthesized in relatively low overall yields (i.e., 3.2 percent), suggesting that new synthetic procedures for the usage of completely conjugated macromolecules in real-world electronic applications are required.

Börner and colleagues described a series of investigations in which they employed a variety of synthetic, sequence-defined peptidomimetics for medication delivery. Combinatorial techniques developed from a single-bead single-compound notion were used to prepare potential candidates from a wide library.

Further examination of the drug-interacting oligomers indicated a considerable dependence on the sequence of the oligomers for both drug-payload capacities and release kinetics. According to a recent study, peptidomimetic synthetic macromolecules may encapsulate 40% more of the same therapeutic payload and have a better sensitivity to release kinetics by merely changing the side chains.

In a similar investigation, a synthesized macromolecule was demonstrated to have a greater drug loading capacity (69%) than the native peptide whilst keeping identical characteristics. In the capacity to uniformly enclose dye molecules, Meier and colleagues also showed homogeneous star-shaped block-macromolecules.

6.8. MACROMOLECULES APPLIED TO PHARMACEUTICAL CHEMISTRY

Macromolecular and polymer research has advanced greatly in recent years, with notable developments in several fields such as polymeric pharmaceuticals, self-assembly systems, implant materials, drug delivery systems, and controlled drug release. These fields are now well established in the sphere of transdisciplinary technology and science.

A growing number of macromolecule applications have necessitated the covalent attachment of polymers to a variety of substrates, including low-molecular-weight medicines, affinity ligands, proteins, oligonucleotides, micro- and nanoparticles. For example, a multidisciplinary Special Issue of Molecules included ten essays from famous scientists who have begun research initiatives in the interesting field of the application of macromolecules in pharmaceutical chemistry. Coviello et al., discusses the application of scleroglucan and certain derivatives in pharmaceutics, namely the production of modified-release dosage forms.

The authors describe a new hydrogel made from this polysaccharide and borate ions, as well as the processes involved in the drug release from the tested dosage forms. Eliyahu et al., highlighted current advances in nucleic acid transport and its various applications in fundamental research, biotechnology, and medicine in the following article.

Non-viral gene delivery vectors, often known as "self-assembled" systems, are based on cationic compounds that spontaneously form complexes with negatively charged nucleic acids. A transition from *in vitro* to *in vivo* gene transfer is also described, with an emphasis on the challenges to effective *in vivo* transfection. Delie and MBlanco- discussed current advances in the design and manufacture of polymeric oral formulations.

Peptides, proteins, and nucleic acids are examples of incredibly powerful new chemicals created via biotechnology. Their oral bioavailability remains very low due to their vulnerability to chemical and enzymatic degradation, as well as inadequate cellular absorption. The most prevalent production processes for polymeric particles are covered, as well as the physiology of particle absorption from the GI tract and the application of polymeric particulate systems to increase insulin oral absorption.

Aulenta et al., consider enzymes to be important instruments in organic synthesis because they can catalyze a wide range of selective chemical changes. However, the utilization of these natural catalysts in the synthesis and post-synthesis modification of dendrimers and hyperbranched molecules is a chemical application that has yet to be thoroughly investigated.

Two hydrolytic enzymes, a lipase from *Candida cylindracea* and a cutinase from *Fusarium Solani pisi*, were tested for their ability to selectively cleave ester groups on the peripheral layer of two families of branched polyamides. Quesnel and Hildgen discussed the relevance of biodegradable polymers in drug carriers and controlled release systems.

The authors suggested a fast-synthetic approach for a polyester-polyethylene multiblock copolymer in which the polyester blocks alternate with polyethylene oxide blocks in a repeated pattern. The copolymers demonstrated here have the fundamental properties necessary for medication microencapsulation.

Following that, Nadeau and Hildgen investigated the various ways of the synthesis of new biodegradable charged polymers for usage in DNA complexation for genetic delivery in various disorders. AFM is also regarded as a strong instrument for semi-quantitative and qualitative studies of non-oriented or orientated polymer film molecular shape and distribution.

Elvira et al., investigated polymer-drugs conjugates utilized as drug delivery systems (DDS), paying special attention to their chemical conjugation. This type of DDS is classified based on the reactive groups' conjugation locations (i.e., via end groups or pendant polymer groups).

The benefits and drawbacks of these DDS are highlighted using instances of recently produced polymer-drug and polymer-protein conjugates. Irache et al., provided a complete review of biodegradable nanoparticles with bio-adhesive capabilities for the oral distribution of poorly accessible medicines.

Orally given nanoparticles can interact with the gastrointestinal surface and form adhesive connections with various mucosal components. The bio-adhesive potential of fluorescently tagged Gantrez nanoparticles with rhodamine B isothiocyanate is summarized. Park and Park provided an overview of the use of biodegradable polymers in drug delivery systems.

The majority of biodegradable polymers have been employed in the form of microparticles, from which the integrated medication is gradually released into the environment. This study examines both traditional and emerging approaches for medication microencapsulation utilizing biodegradable polymers. Furthermore, the properties and degradation behavior of biodegradable polymers currently employed in medication delivery are reviewed.

Moreover, Shibata et al., display their work on the enhancement of protein treatments by polymer-conjugation as an effectual DDS, along with their protein-drug system intended to promote disease proteomic-based drug research for protein therapies and conquer the clinical challenges of using proteins as effective and safe drugs due to low stabilization and pleiotropic actions *in vivo*.

6.9. MACROMOLECULAR TECHNOLOGIES: APPLICATIONS AND IMPROVEMENTS

This seminar provided, as in past years, an excellent opportunity to learn both about new technology and freshly created improvements that improve established research methodologies. A frequent meeting highlight is the honoring of an exceptional contribution to technological growth.

Csaba Horvath (Yale University, New Haven, USA) was honored this year for his contributions to the evolution of contemporary chromatography. The plenary speeches served as an ideal backdrop for demonstrating how basic science drives the discovery and development of the numerous research methodologies and technologies that were covered in depth throughout the smaller concurrent sessions.

Ronald Evans (Salk Institute, La Jolla, USA) discussed the complexities of nuclear hormone receptor activity and highlighted potential drug-drug interactions. He highlighted an intriguing adaption process that allows the body to increase resistance to a given chemical, as well as how this xenobiotic response' aids in detoxification and elimination of the chemical from the body.

This reaction, once initiated, eliminates a range of chemicals from the body and can be triggered by components found in non-prescription compounds (such as St John's Wort). Upon stimulation, it eliminates a wide range of chemicals from the body.

The active element in birth control pills, for example, may be cleared from the body by the xenobiotic reaction, offering a scientific explanation for many miracle cures, like protease inhibitors, which are then used to treat HIV.

Roger Brent (Molecular Sciences Institute, Berkeley, USA) presented the creation of computer software that predicts how a biological system would react to a given input. Brent believes that realistic modeling of cellular function will be achievable in the future because of developments in computing power, computational methodologies, and biological understanding.

As an example, the group is creating datasets from cells that have been exposed to varying levels of a signal (such as yeast mating pheromone). The expression of different fluorescent protein constructs offers information about the active promoters and allows for the measurement of the biological reaction to the signal.

Andrew Marks (Columbia University, New York, USA) revealed a sequence of important protein-protein interactions between proteins in heart muscle that build and control calcium-release channels. Large cytoplasmic domains in these intracellular (sarcoplasmic reticulum) channels function as scaffolds for the extra proteins that control pore shape and channel activity.

Heart failure can result if such connections are disrupted by hyperphosphorylation. As a result, the proteins implicated in these interactions are possible treatment targets.

Lehrer discussed protegrins, an intriguing family of broad-spectrum antibiotics that rapidly rupture the bacterial outer membrane, killing cells in minutes. Kornberg published a three-dimensional structure of the *Saccharomyces cerevisiae* RNA polymerase II transcription machinery, which comprises roughly 50 polypeptides.

Finally, Wolffe outlined an approach for employing designed zinc-finger proteins as transcription factors to activate or repress certain genes, which is an interesting development with both medical and biotechnological implications. ABRF research groups perform studies to evaluate and compare the core facilities supplied by member laboratories.

6.10. EXPERIENCES WITH APPLICATIONS OF MACROMOLECULAR TOOLS IN SUPRAMOLECULAR CRYSTALLOGRAPHY

Even though the underlying principles of macromolecular (also known as protein) crystallography and small-molecule crystallography (here concerned primarily with organic molecules) are largely the same, the experimental and analytical initiatives in such two categories are quite clearly different, partially due to tradition and partly due to actual idiosyncrasies of these domains.

This is evident in the typically simple crystallization methods used in small-molecule crystallography and the highly developed, typically high-throughput, miniaturized and robotized macromolecular crystallization techniques, as well as the fact that almost all organic crystal structures are rectified automatically and routinely by direct methods, whereas this method is utterly irrelevant for typical-size macromolecular structures.

A significant difference involves the possible resolution, which is nearly always quite high with tiny molecules. Even when restricted by the wavelength of Cu K radiation, and in protein, crystallography is still rare even at the nominal atomic resolution of 1.2 established by Sheldrick1, as seen by the consistent level of 2% of such structures in the Protein Data Bank.

For example, although protein crystallographers always base their models on electron density maps, small-molecule crystallographers can operate fairly easily using atom/peak lists derived by computer interpretation of such maps. There are, nevertheless, clear lines of convergence.

The loop approach, for example, developed for mounting protein crystals for cryogenic experiments4 is gaining popularity in small-molecule crystallography, and both communities employ cryogenic temperatures for routine data collecting (although the reasons in the two cases may be somewhat different).

Furthermore, the high-resolution barrier is rapidly being broken down, with record-breaking ultrahigh-resolution structures of proteins (0.48) and nucleic acids (0.55) being available in the PDB. The prevalent SHELX system of crystallographic programs, first created for tiny molecules and then very successfully adapted by its inventor to a flexible system currently extensively used in macromolecular crystallography, is a very encouraging example of convergence.

Generally, the computational tools in the two categories are highly different, with small-molecule programs incapable of handling the massive macromolecular instances and macromolecular programs frequently hard-wired for the structural features of biopolymers.

The sophisticated computational tools of protein crystallography could become immensely useful for smaller molecule crystallography too though, particularly whenever the "small molecules" are not really tiny at all, just like in self-assembling supramolecular structures. Researchers demonstrate that employing "routine" macromolecular techniques may considerably aid in the resolution of crystal structures of supramolecular assemblies that cannot be resolved using "regular" small molecule crystallography methods.

6.11. APPLICATIONS OF SURFACE-GRAFTED MACROMOLECULES

The polymerization of monomers with specified characteristics is arguably the most basic way of the production of specialized polymers intended for specific uses. In many circumstances, this design philosophy will show to be the most efficient at obtaining the desired result.

Given the enormous number of monomers easily accessible from chemical manufacturers today, as well as the various synthetic methods developed over the last century that facilitate the integration of a wide range of functionalities into a single macromolecule, and the need for supplemental chemical reactions which adjust macromolecules after polymerization may not have been obvious.

Besides the ease of the direct synthesis technique, multiple issues may develop fast at different phases along the synthetic route. Firstly, the monomer could be incompatible with the reaction conditions necessary to include a given functionality. In this situation, time and energy would be expended on what could become such an immensely complex monomer synthesis.

Moreover, the yield of altered monomer may be too tiny and too expensive to make significant amounts of polymer. In the case of successful monomer synthesis, the actual polymerization may fail to yield reasonably high molecular weights or low poly-dispersities.

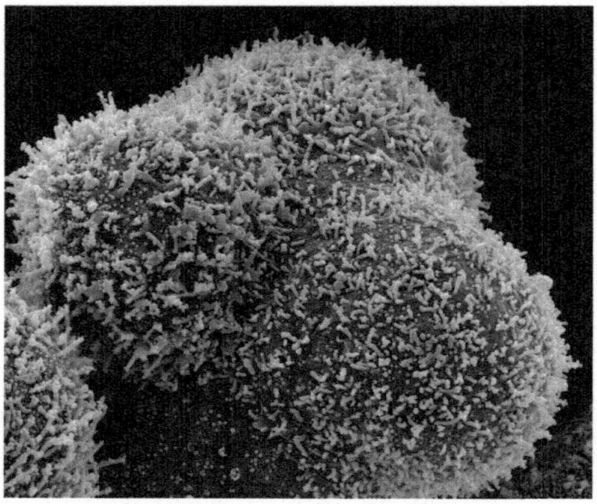

Figure 6.4. Novel Coronavirus SARS-CoV-2.

Source: Image by Flickr

Several current polymerization processes can produce polymers that fit both of these criteria, but they cannot handle every monomer or those with bulky pendant groups, such as bioconjugates. It is possible that the functional groups introduced into the initial monomer are not suitable for the polymerization reaction conditions (i.e., temperature, solvent, etc.).

In circumstances when it is possible to protect sensitive chemical groups, the deprotection reaction may not complete, invalidating the value of beginning with the modified monomer in the first instance. These (and other unspecified) difficulties apply not just to bulk polymerization operations, but also to the synthesis of macromolecules with at least one of their points attached to a substrate.

The entropic restrictions associated with such confinement present another barrier that may impede the direct proliferation of a given monomer via surface-initiated polymerizations. With these considerations in mind, there is a need for an alternate method of creating functionalized polymer chains is apparent.

The PPM of macromolecules tethered to flat impermeable surfaces is summarized, as are a few uses of such systems. PPM's capacity to change the chemical nature and structure of a polymer is combined with the ability to control the degree of chemical modification and spatial distribution of the recently introduced modifier along the macromolecule.

Whereas the degree of "chemical coloring" is controlled by modifying the reaction conditions (temperature, chemical modifier concentration, coupling chemistry, catalyst, and solvent type), the co-monomer distribution of the parent and new monomers is controlled by changing the "degree of confinement" of the parent polymer by either making adjustments the solvent quality or varying the grafting density of the macromolecular graft on the substrate.

Complicated brush systems with the position-dependent distribution of chemical modifiers that may be used to exhibit orthogonal gradients in functionality are feasible. Substrates with two unique chemical patterns of varying shapes and diameters separated by sharp boundaries can also be created.

Copolymers with tunable co-monomer distributions that span the di-block-random distribution spectrum display responsive properties. These can be used in a wide range of applications that depend just on the stimuli-responsive nature of surfaces.

6.12. INDUSTRIAL APPLICATIONS OF MACROMOLECULES

There seem to be three major kinds of macromolecules that are significant in industry, in addition to the highly important biologic macromolecules (proteins, lipids, polysaccharides, and nucleic acids). Elastomers, fibers, and plastics are examples of these materials.

Elastomers are elastic and extremely flexible macromolecules. Because of their elasticity, such materials may be utilized in items such as elastic waistbands and hair bands. Such items can be stretched, but they always return to their original position. Rubber is a natural, non-man-made elastomer.

You most likely wear fiber macromolecules. Polyester, nylon, and acrylic fibers are utilized in a variety of products, including shoes, belts, and shirts and blouses. Fiber macromolecules are string-like molecules that, when weaved together, are extremely durable. Silk, cotton, wool, and wood are examples of natural fibers.

What did humans do prior to the invention of plastics? These macromolecules are found in many of the products we use every day. There are numerous varieties of plastics, and they're all created through a procedure process called polymerization, which involves the combining of monomer units to produce the plastic polymers. Until lately, all plastics were derived from petroleum.

Figure 6.5. Industrial Landscape.

Source: Image by Flickr

Bioplastics are a novel class of polymers that have the benefit of being made from renewable resources. Bioplastics, unlike ordinary plastics, may break down or decay, making them safer for the environment. People who care about the environment recycle plastic products. There are no naturally occurring plastics.

6.13. ANTIOXIDATIVE BIOMACROMOLECULES

6.13.1. Proteins

Proteins are the building blocks of life and vital components of cells and tissues of living beings, where they play crucial roles in a variety of biological and physiological processes. Antioxidant activity, for example, is an important representative.

Some antioxidative proteins, such as SOD, Prxs, and GPxs, can help the body maintain its redox state by inhibiting intracellular ROS generation or encouraging ROS removal. Extracellular SOD3, cytoplasmic SOD1, and mitochondrial SOD2 are the three kinds of SOD.

These enzymes can convert superoxide radicals into harmless compounds. Prxs are a thioredoxin-dependent peroxidase family with six subtypes found in mammals, all of which are expressed mostly in cells except for GPx3.

Prxs have the ability to catalyze the redox reaction of intracellular H_2O_2. To defend cells from oxidative stress, peroxy-nitrite, and organic oxides are used.

Non-enzymatic metal chelating proteins such as transferrin, trans-calcin, and seleno-protein tightly regulate metal ion levels and avoid excessive buildup in the body. Furthermore, DNA repair enzymes and proteolytic enzymes are in charge of addressing free radical-induced cell damage and reconstructing damaged cell membranes.

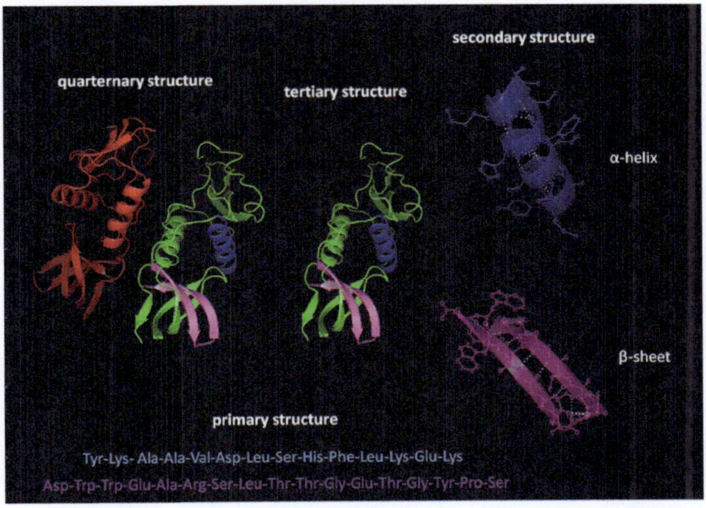

Figure 6.6. Levels of structural organization of proteins.

Source: Image by Wikimedia commons

Furthermore, these enzymes can aid in the identification, detoxification, decomposition, and elimination of oxidized or damaged macromolecules such as proteins, DNA, and lipids, therefore preventing accumulating harm to the body.DNA polymerases, glucoamylases, nucleases, proteases, proteasomes, and peptidases are the most common enzymes.

These are found in the cytoplasm and mitochondria of mammals and are crucial components of the body's antioxidant defense mechanism. Furthermore, several antioxidant metalloproteins participate in a variety of brain functioning processes.

For example, selenoproteins are recognized with selenocysteine at the catalytic site, and the majority of them are expressed in the cerebral cortex and hippocampus neurons, where they safeguard neuronal systems. The selenoprotein family includes glutathione peroxidase I-IV and VI (GPx1-4, 6), thioredoxin reductase I-III (TrxR1-3), and methionine sulfoxide reductase B (MsrB).

6.13.2. Polypeptides

It has been claimed that certain peptides generated by animals and plants can function as nutrition for human growth and development. Meanwhile, they can provide important physiological functions to the human body, such as neural, hormonal, and immunological regulation, as well as antithrombotic, antihypertensive, anti-cholesterol, anti-bacterial, anti-viral, anti-cancer, antioxidation, anti-aging, and other physiological activities, the most essential of which is antioxidation. Despite an insufficient knowledge of antioxidative peptides' antioxidative action, massive investigations have revealed that these peptides may scavenge free radicals and metal transition chelates or reduce lipid peroxidation. These peptides' antioxidative properties are connected to their structural characteristics (amino acid compositions, molecular size, hydrophobicity, etc.).

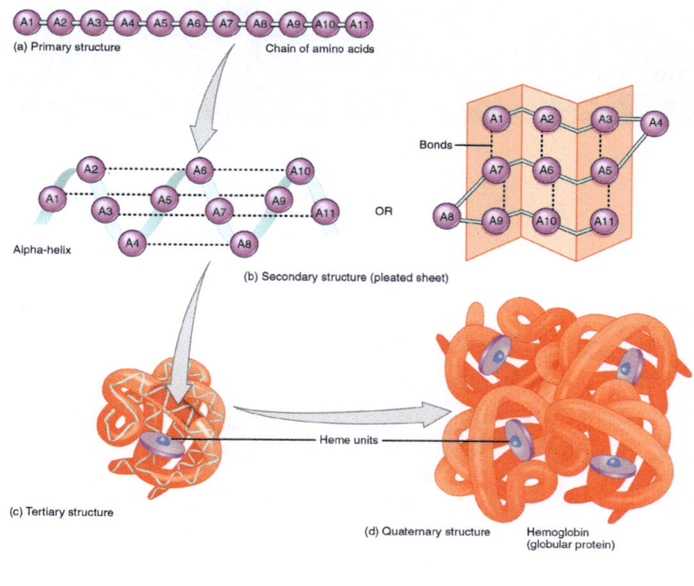

Figure 6.7. Image showing peptide bonds.

Source: Image by Wikimedia commons

This has been proposed that hydrophobic amino acids near the N-terminus of the peptide chain can boost the peptide's antioxidative action by increasing the affinity between the peptide chain and hydrophobic cell targets. Some peptides are also involved in signal transduction, which protects the organism from oxidative damage by triggering the transcription of genes encoding antioxidative enzymes. For example, Orsini Delgado et al. discovered that several natural antioxidative peptides found in the seeds of Amaranthus plants may scavenge free radicals and block linoleic acid oxidation. These antioxidative peptides are also present in numerous proteins, including albumin, globulin, and mucin.

6.13.3. Glycoproteins

Glycoproteins (GPs) are glycoconjugates made of a polypeptide backbone to which one or more carbohydrate units are attached via covalent bonds, often N- and O-glycosidic linkages. GPs are found in many plants, as well as mammals, fungi, molds, and algae. Many biological actions of GPs are active, including anticancer, antioxidant, antiviral, and blood glucose lowering.

GPs are also involved in the preservation of protein structure and stability, the control of protein processing and transport, cell adhesion, and cell surface or intracellular recognition. The biological activities of GPs are influenced by glycosylation sites, amino acid sequences, monosaccharide content, and glycoprotein linkage.

The rapid advancement of biotechnology encourages substantial research into the link between the biological activity, function, and structure of GPs. Dai et al., for example, used the Box-Behnken design to discover the best conditions for extracting a GP (SPMG) from the muscle of Pharaoh Cuttlefish.

The crude extract was isolated and purified into two fractions, SPMG-I and SPMG-II, using DEAE-Nevis 52 ion exchange column chromatography and Sephadex G-100 gel chromatography, with molecular weights of 42.5 and 36.3 KDa, respectively.

Both fractions were high in Glu, Asp, Lue, Arg, and Lys, with glucose being the most abundant monosaccharide molecule. SPMG-I was formed by O-glycosidic bonds, whereas SPMG-II was formed through a -elimination process. SPMG, SPMG-I, and SPMG-II all had substantial antioxidative activity, with SPMG-I having the greatest activity.

Furthermore, Li et al. identified and purified an antioxidative GP with a molecular weight of 27.2 KDa from the ethanol-soluble protein fraction of *Mustelus griseus*. This GP yielded a total of 17 amino acids, with serine being the most abundant. Trehalose, arabinose, galactose, glucose, and mannose have a 1.00:1.53:7.27:9.07:2.09 ratio.

The resulting GP structure was characterized by polysaccharide and protein characteristics coupled by N- and O-glycosidic linkages clearance of 1,1-diphenyl-2-picrylhydrazol (DPPH) by this GP achieved 96.732.33%, which was greater than the clearance of ascorbic acid at 5.0 mg/ml.

6.14. APPLICATIONS OF ANTIOXIDATIVE BIOMACROMOLECULES

Antioxidant biomacromolecules have been shown to efficiently decrease oxidative stress and prevent the development of a variety of chronic illnesses. Such macromolecules are indeed an excellent source for biomedicine, functional foods, and cosmetics product development since they are non-toxic and pose no risk to human health.

6.14.1. Biomedicine

ROS and oxidative stress rise with the formation of malignant tumors and myeloproliferative diseases. As a result, it is required to enhance redox equilibrium by changing the production of associated antioxidative proteins. Antioxidant properties of bioactive macromolecules are critical for maintaining the body's redox state and decreasing the damage caused by illnesses or medications.

Phyllanthus niruri was used by Bhattacharyya et al. to identify an antioxidative protein. The isolated protein, through activating the PI3k/Akt signaling pathway, may decrease liver weight loss, increase phosphorylation of the transcription factor p65 subunit, and ameliorate aspirin-induced oxidative damage and apoptosis, therefore reducing the detrimental effects of aspirin.

Meanwhile, Sun et al. discovered a physiologically active polypeptide from the mycelium of *Pleurotus eryngii*. As per the findings, at a concentration of 0.05-2 mg/ml, the polypeptide might inhibit tumor cell proliferation, promote macrophage proliferation, the release of tumor necrosis factor (TNF)- and interleukin 6 (IL-6), and the expression of TLR2 and TLR4, and enhance macrophage phagocytic capacity via NO and H_2O_2.

The discovered polypeptide could efficiently decrease DPPH, O_2^-, and OH^- at concentrations ranging from 0.2-1mg/ml. This polypeptide's antioxidative and immunostimulatory properties make it a promising physiologically active option for antiviral medicines. The PARK7 gene encodes the DJ-1 protein, which is a redox protein with numerous biological roles.

When exposed to oxidative stress, it becomes active and acts as a transcriptional regulator of antioxidative proteins. In their work, Trivedi et al. found that when Caki-2 cells were treated with cisplatin, ROS continued to build in dosage and time-dependent ways, finally leading to cell death. A proteomic study revealed that DJ-1 protein expression steadily reduced in the presence of Caki-2 cell death.

When DJ-1 was overexpressed, cell apoptosis was inhibited, but it accelerated when the PARK7 gene was taken out. These studies demonstrated that the DJ-1 protein can remove ROS from kidney cells and alleviate cellular oxidative stress. Nevertheless, this is not true that increased DJ-1 expression is always linked with a meaningful impact.

The rationale for this is because, in addition to protecting transformed cells from increasing ROS, the redox adaption mechanism of malignant cells can promote reduced cell death, greater DNA repair ability, and enhanced drug resistance.

Kim et al. discovered a bidirectional change in DJ-1 expression in a mouse model of mastocytosis, i.e., DJ-1 was degraded and lowered owing to reduced ROS in animals with mild mastocytosis but considerably elevated in a malignant mouse type of mastocytosis.

During the latter situation, IL-6 triggered PARK7 gene transcription, which increased DJ-1 expression to compensate for DJ-1 loss caused by oxidation and reduce oxidative stress, creating a favorable physiological environment for mast cell proliferation.

While using an anti-IL-6 antibody prevented IL6 receptor signals, it suppressed ROS rise and DJ-1 production while also significantly reducing the number of mast cells in tissue and blood. As a result, reducing DJ-1 expression by inhibiting the IL-6 signal may be an effective adjuvant therapy for individuals with advanced mastocytosis. Antioxidant biological macromolecules are found in a broad variety of higher plants, fungus, molds, algae, and bacteria. Extraction, separation, and purification methods are critical for preserving these macromolecules' biological activity.

6.14.2. Functional Foods

Traditional Chinese medicinal ingredients rich in diverse bioactive macromolecules are plentiful in China and can serve as key raw materials for the formation of functional foods. A polysaccharide was isolated from *Gynostemma pentaphyllum* by Wang et al. *In vitro* experiments revealed that this polysaccharide inhibited the elimination of DPPH free radicals, superoxide anions, and ABTS free radicals in a dose-dependent way.

Meanwhile, animal studies found that this polysaccharide successfully lowered body weight, fasting blood glucose, ALPA, ALT, AST, and BUN levels in the blood. Because of their particular living environment, antioxidative macromolecules from marine species have specific structures and are active in controlling the immunological function of the human body.

As a consequence, the bioactive activities of marine creatures are gradually being identified and used in the synthesis and manufacture of functional meals. Qi et al. isolated sea cucumber polysaccharide (PESCPL) from sea cucumber processing liquid as a raw material using proteolysis and electroosmosis. PESCPL is mostly made up of mannose, with a little glucose and fucose thrown in for good measure.

PESCPL may efficiently eliminate DPPH, hydroxyl free radicals, and superoxide anion free radicals, as well as boost the activities of catalase and SOD, decrease serum malondialdehyde, cholesterol, and triglyceride levels, and raise the content of high-density lipoprotein cholesterol. As a result, it is a naturally occurring antioxidant that may be utilized as a dietary supplement for dyslipidemia.

Furthermore, various macromolecules possess antioxidative properties. The projected effects of these compounds, though, are modest due to the relatively low bioavailability and instability in the gastrointestinal environment following oral delivery.

According to research, when bioactive macromolecules and polypeptides are synthesized into nanoparticles using nanotechnology, this problem may be efficiently overcome. Tang et al. created self-assembled chitosan/poly (-glutamic acid) (-PGA) nanoparticles (CS/-PGA). Meanwhile, the discovered nanoparticles were revealed to efficiently scavenge free radicals and reversibly loosen the tight link between Caco-2 cells to increase catechin distribution across Caco-2 monolayers. The nanoparticle preparation approach described above generates fresh concepts for the synthesis of innovative peptide-based antioxidant functional meals.

6.14.3. Skincare Products

Antioxidant properties of biological macromolecules can efficiently eliminate ROS and free radicals to preserve skin cells, reduce aging, and limit melanin synthesis.

Chen et al. discovered that squid ink polysaccharides substantially decreased oxidative damage to fibroblasts mediated by increased NADPH oxidase and connexin 43, and prevented ROS-induced upregulation of matrix metalloproteinase (MMP) 1 and MMP9, reducing MMP9-mediated skin aging.

Hou et al. isolated collagen peptides from cod skin and discovered that they can boost antioxidant activity, minimize water and fat loss, repair endogenous collagen and elastin fibers, and preserve the type III to type I collagen ratio.

Moreover, with good moisture absorption and retention ability, the extracted collagen polypeptide can minimize skin damage induced by UV radiation exposure. To coat fish oil, Wang et al. created a conjugate of mulberry polysaccharide and whey protein.

The produced fish oil emulsion with reduced particle size enhanced the emulsification capacity and stability of whey protein, resulting in a considerable improvement in the antioxidative activity of fish oil. As a result, antioxidant peptides are gaining popularity in the skincare sector.

6.14.4. Other Bio-Products

Aside from being used as a component in biomedical, cosmetics, and functional foods, antioxidant biological macromolecules in food and medication packaging materials have a wide range of applications as people strive for a higher quality of life and a greater awareness of environmental conservation.

As a result, it has been the research and development focus for novel food packaging materials that are safe, non-toxic, bioactive, biodegradable, and recyclable. A number of approaches may be used to create novel functional materials with biological activity from antioxidant bioactive macromolecules with unique structures, which will be the raw ingredients for the development of food packaging materials in the future.

Spizzirri et al. used a free radical-induced grafting process to create conjugates of catechin-alginate and catechin-inulin. The functional materials demonstrated high antioxidant activity, which might be highly valuable in optimizing food preservation. This strategy, as described before, gives inspiration for developing novel packaging materials.

Antioxidative bioactive macromolecules have varied shapes, as well as the manufacturing techniques have a substantial impact on their biological activity. Various preparation procedures have their advantages and disadvantages, as well as their scope of applicability.

To maintain the bioactivities, appropriate procedures must be established depending on the physicochemical features or bioactivities of the target molecules. Furthermore, with ongoing technological improvement, separation and purification methods of bioactive macromolecules must be improved.

It is predicted that by doing so, it will be possible to investigate the roles of more bioactive macromolecules and hasten their use in biomedicine, clinical diagnosis and treatment, and functional foods, finally promoting the faster and more efficient growth of human health.

6.15. CONCLUSION

The interest in solid phase synthesis is driven by the ease of purification, however, it shows some drawbacks too. Polymers are nowadays present everywhere in our everyday life. One of the most widely spread methods is solid phase synthesis. Solution phase synthesis approaches are therefore preferred by some research groups.

Since the implementation of the first fully synthetic plastic by Leo Baekeland in 1907, this area of research has extensively grown. As a consequence, the field of "sequence-defined polymers" has been created. However, the 21st century has come with new challenges and new areas of research have emerged such as sequence-controlled polymers.

This term has been used to describe polymers in which the monomer sequence is regulated and the dispersity is reduced. New polymers have been created and their performances were continuously improved. For example, the chain length can be limited because of incomplete coupling steps or aggregation of the oligomers, thus limiting the accessibility of the chain ends, etc.

REFERENCES

1. Aksakal, R., Mertens, C., Soete, M., Badi, N. and Du Prez, F., 2021. Applications of Discrete Synthetic Macromolecules in Life and Materials Science: Recent and Future Trends. *Advanced Science*, [online] 8(6), p.2004038. Available at: <https://onlinelibrary.wiley.com/doi/10.1002/advs.202004038> [Accessed 21 June 2022].
2. Casey J. Galvin and Jan Genzer, 2011. *Applications of surface-grafted macromolecules derived from post-polymerization modification reactions*. [online] People.engr.ncsu.edu. Available at: <https://people.engr.ncsu.edu/jgenzer/pubs/pub-12-13.pdf> [Accessed 21 June 2022].
3. Chen, M., Zhang, G., Chen, S., Zhang, X. and Chen, W., 2019. Preparation and application of macromolecules with a fluorescence effect in polymer processing. *Designed Monomers and Polymers*, [online] 22(1), pp.187-198. Available at: <https://www.tandfonline.com/doi/full/10.1080/15685551.2019.1687082> [Accessed 21 June 2022].
4. Hardin, S., 2001. Macromolecular technologies: applications and improvements. *Genome Biology*, [online] 2(5), pp.reports4012.1. Available at: <https://www.ncbi.nlm.nih.gov/pmc/articles/PMC138932/> [Accessed 21 June 2022].
5. Salomon, C., 2005. Macromolecules Applied to Pharmaceutical Chemistry. *Molecules*, [online] 10(1), pp.3-5. Available at: <https://www.ncbi.nlm.nih.gov/pmc/articles/PMC6147525/> [Accessed 21 June 2022].
6. Song, R., Wu, Q., Yun, Z. and Zhao, L., 2020. Advances in Antioxidative Bioactive Macromolecules. *IOP Conference Series: Earth and Environmental Science*, [online] 512(1), p.012094. Available at: <https://iopscience.iop.org/article/10.1088/1755-1315/512/1/012094/pdf> [Accessed 21 June 2022].

CHAPTER 7

Solid State Macromolecules

CONTENTS

To begin, despite being a specialty discipline, polymer science is a meeting place for almost all-natural sciences in terms of scope of applicability and methodology employed in its application. This is followed by a wide taxonomy of polymeric materials in the lecture.

7.1. INTRODUCTION

In this latter setting, distinctions are drawn between linear chain molecules and networks, flexible and stiff molecules, and macromolecules with and without a periodic molecular structure. Throughout this categorization, the focus is directed to the following specific features that distinguish macromolecules:

The broad connection by valence bonds can have crucial implications for material science, such as the inherent capacity of macromolecules to give birth to materials of remarkable stiffness and strength if these bonds are all aligned, indicating that the molecules are orientated.

In the case of random molecular states and chain flexibility, the large range of conformational possibilities gives birth to a unique state of matter characterized by unusually high elastic extensibility, which is entropic in nature (rubber elasticity). If the chains are stiff, a distinct liquid crystalline state may form.

If the macromolecule is made up of many chemically diverse monomeric elements that are also in an aperiodic sequence, an immense range of such sequential possibilities emerge, making them ideal for storing information and nuanced management of complicated chemical processes. This is why macromolecules are the primary elements of living organisms, holding and transferring genetic information and controlling metabolic activities.

In the case of periodic structures, valence bond continuity can be the source of unique electronic characteristics, which is a recent area of interest.

Macromolecular substances react to radiation in an unusual way because, due to their high molecular scale, a single event within one molecule influences a significant amount of material. Specific examples are used in the presentation to demonstrate the effects of factors.

These have been chosen to correspond to issues and approaches in several disciplines of physics, spanning from mechanics to superconductivity, to biophysics, and particle physics to crystal formation. After setting the stage, the remainder of the content focuses on flexible chain molecules with precisely periodic structures.

7.2. PROPERTIES OF SOLIDS

Solid molecules do not move in the same way as liquid or gas molecules do. Instead of moving, solid molecules merely vibrate and spin in situ. Ionic or strong covalent bonding holds solids together, and the attractive forces between atoms, ions, or molecules in solids are very strong.

Indeed, these pressures are so powerful that particles in a solid are kept in fixed places with minimal flexibility of movement. Solids have distinct forms and volumes and are not compressible in any way.

Solids are classified into two types: crystalline solids and amorphous solids.

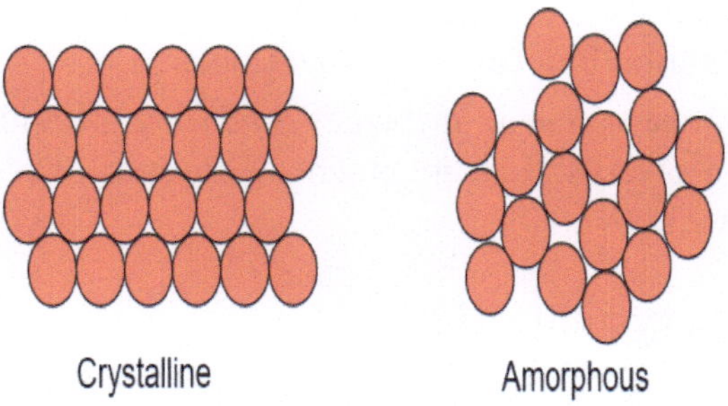

Figure 7.1. Particles in crystalline and amorphous solids.

Source: Image by Wikimedia Commons

Crystalline solids are those that have their atoms, ions, or molecules arranged in a regular, well-defined pattern. The unit cell is the smallest repeating pattern of crystalline solids, and unit cells are similar to bricks in a building in that they are all identical and repeating.

Amorphous solids are the other major form of solid. The structures of amorphous solids are disordered. Despite the fact that their molecules are close together and have minimal freedom of movement, they are not ordered in a regular sequence like those in crystalline solids. Glass and plastics are common examples of this type of solid.

7.3. ORGANIZATION IN THE SOLID STATE: CRYSTALLINITY

The periodicity needed by a crystal lattice in periodic molecules is given, in one direction, by the periodicity inside the molecule itself. There are examples of 'crystal structure' as described by the locations of the atoms within the unit cell.

However, the 'crystal structure' alone is insufficient for understanding the nature of crystalline polymers since the question of what happens to the chain as a whole emerges. Three kinds of crystals can be described in this context based on their various ways of formation. The latter is associated with crystallization:

1) Concurrent with chain expansion ('nascent crystallization').
2) From previously produced chains in a random state ('conventional crystallization').
3) As it is in 2, except the chains are expanded before crystallization ('orientation induced crystallization').

In literature, examples of instances 1)-3) are provided as each lead to a distinct form of crystal arrangement, with case 2) receiving the most attention and space).

7.3.1. Nascent Crystallization

Professor Wegner at Freiburg pioneered and pursued a unique area of this vast field. This is the transformation of a macroscopic monomer crystal into a macroscopic polymer crystal by connecting (polymerizing) the monomeric parts to create chains while keeping the macroscopic crystal entity, which now consists of chains in the crystallographic register throughout.

Such polymer macro crystals occur only in rare circumstances, however, whenever they do, they constitute the nearest contact point between polymer research and classical solid-state physics, with all the novelty that covalent bonding along one direction of the crystal involves.

7.3.2. Conventional Crystallization

The study of the crystalline polymeric solid state revolves around the crystallization of flexible polymer chains from solution or melt. The most unique and surprising characteristic of this crystallization is that the fundamental crystal entity is an electron microscopic lamella inside which the chains are folded in a repeating fashion, giving birth to layers and specifying layer thickness.

The folded length, and hence the layer thickness, is not constant but is influenced by the crystallization circumstances in a very precise fashion (undercooling in particular). This chain folding is a fundamental topic in solid polymer physics and has been the focus of extensive theoretical and experimental investigations.

The ideas seek to account for folding, as well as the consequent fold length and crystal growth rate Agreement with experiments may be highly satisfying, especially when compared to the large disparities that can occur in comparable endeavors in the field of crystal development of simple compounds, where chain folding is not even considered and the difficulties are restricted to growth rates alone.

Chain folding has numerous fascinating structural ramifications, such as sectorization, as proven by, another rare occurrence in crystalline materials. The nature of the folds themselves has been extensively examined and argued, particularly how far they are regular in the sense.

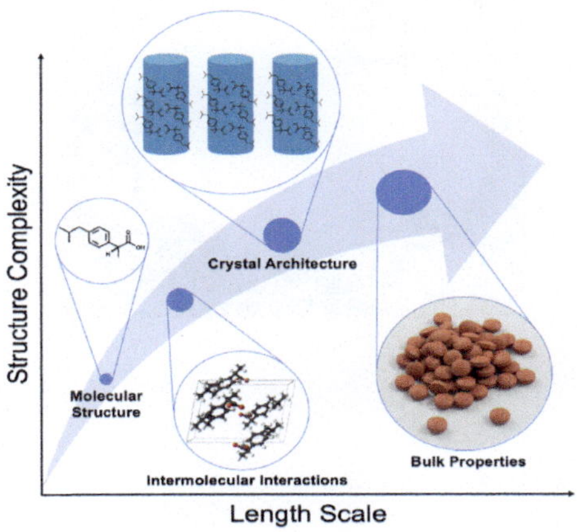

Figure 7.2. Complexity and Scale of Structures within Crystalline Materials.

Source: Image by Wikimedia Commons

Regardless of the final solution to this dilemma is that the chains fold, and the length of the fold defines the lamellar thickness. Furthermore, crystalline polymers are known to include a certain quantity of disordered amorphous matter, which must be accommodated.

The model gives one, but by no means unique, approach to account for this architecturally, whereas many structural regularities found (e.g., the described sectorization) are more easily interpretable if a sufficient level of regularity is attributed to the fold structure.

The chain folded lamellae are simply the building blocks for bigger-scale structural groupings that make up a macroscopic piece of crystalline thermoplastic. The key lesson is that the organization of a crystalline polymer is hierarchical, with various organizational principles applied at different diameters stages.

7.3.3. Orientation Induced Crystallization

Crystallization is aided by the stretching of a flexible long chain molecule. Even more crucially, the resulting crystals will be fibrous, with the chains basically stretched within the fibrous crystals (as one would intuitively expect from long chains).

Figure 7.3. Fibrous crystals.

Source: Image by Wikimedia Commons

Chain extensibility and chain extension methods are key branches of polymer research in and of themselves, having implications for hydrodynamics and rheology, as well as explicit implications for the production of fibers with suitable strength and stiffness for industrial applications.

Only a percentage of the chains are stretched at any given moment under normal conditions of chain stretching, and only these generate fibrous crystals. According to Professor Pennings' pioneering work at Groningen, the rest of the links that remain unextended will then crystallize in a chain folded platelet shape, with the fibrous crystals currently present acting as nucleation centers.

The resultant organism, known as shish-kebabs,' is composed of a series of parallel chain folded platelets strung on an extended chain type fibrous backbone, reflecting a specific mix of the two morphologies.

Structures of this type, in a variety of shapes and modifications, exist anytime crystallization occurs under stress and during flow, and are thus highly common and of basic and practical importance. To summarize, even when limited to the relatively narrow realm of flexible and rigidly periodic chain molecules, the unexpected variety of organizing options in the field of macromolecules becomes obvious.

The structures that emerge are intriguing in and of themselves, but they also lead to many perplexing physics difficulties and have various real repercussions for macroscopic qualities and, eventually, technology usage.

7.4. THERE ARE FIVE TYPES OF CRYSTALLINE SOLIDS

7.4.1. Ionic Solid

Composed up of positive and negative ions that are kept together by electrostatic forces. These have extremely high melting temperatures, are extremely brittle, and thus are poor solid-state conductors. Table salt, NaCl, is an example of an ionic solid. Ionic compounds crystallize from oppositely charged ions: a positively charged cation and a negatively charged anion. Ionic bonds are difficult to break because of the strong attraction between opposing charges. This means that ionic compounds have extremely high melting points, frequently ranging from 300 to 1,000 degrees Celsius (572 to 1,832 degrees Fahrenheit). Whereas the crystals are stiff, brittle, and nonconductive, most ionic compounds may be dissolved in water to generate a liquid of free ions that conduct electric current. They might be simple binary salts, such as sodium chloride (NaCl) or table salt, wherein one atom of a metallic element (sodium) is bound to one atom of a nonmetallic element (chlorine). These might also be made up of polyatomic ions like NH4NO3 (ammonium nitrate). Polyatomic ions are groupings of atoms that share electrons (through covalent bonding) and operate as if they were a single charged ion in a molecule.

7.4.2. Molecular Solids

Composed of atoms or molecules joined by London dispersion forces, dipole-dipole forces, or hydrogen bonds. They have low melting points, are flexible, and are poor conductors. Sucrose is an example of a molecular solid. Molecular solids are made up of covalently bound molecules that are attracted to one another by electrostatic forces (called van der Waals forces). Because covalent bonding entails the sharing of electrons rather than the direct transfer of those particles, the shared electrons may spend more time in the bigger atom's electron cloud, resulting in weak or fluctuating polarity.

Figure 7.4. Fully grown table sugar (sucrose) crystals.

Source: Image by Wikimedia Commons

Because the electrostatic interaction between the two poles (dipoles) is far less than ionic or covalent bonding, molecular solids are softer and have lower melting temperatures than ionic crystals. (Plenty will melt at temperatures below 100°C (212° F). The majority of molecular solids are nonpolar. Such nonpolar molecular solids do not dissolve in water but do dissolve in nonpolar solvents like benzene and octane. Sugar and other polar molecular solids dissolve rapidly in water. Nonconductive molecular solids. Ice, sugar, halogens such as solid chlorine (Cl_2), and compounds containing a halogen and hydrogen such as hydrogen chloride are instances of molecular solids (HCl). Fullerene "buckyballs" are molecular solids as well.

Figure 7.5. Buckyballs.

Source: Image by Flickr

7.4.3. Covalent-Network (Also Called Atomic) Solids

Covalent bonds unite the atoms; intermolecular forces are also covalent bonds. They are exceedingly hard, have very high melting temperatures, and are poor conductors. Diamond, graphite, and fullerenes are examples of this type of solid. Graphite has just a 2-D hexagonal structure and so is not as hard as diamond. Only weak London forces hold the graphite sheets together!

Individual molecules do not exist in a network solid. Massive crystals are formed when atoms are covalently bound in a continuous network. Each atom in a network solid is covalently connected to every other atom. Ionic solids and network solids have comparable characteristics.

They are extremely hard, brittle solids they are extremely hard, brittle solids with exceptionally high melting points (higher than 1,000°C or 1,800°F). They do not dissolve in water and therefore do not conduct electricity, unlike ionic compounds.

Diamonds, amethysts, and rubies are examples of network solids.

7.4.4. Metallic Solids

Consist of metal atoms linked together by metallic bonds. They have high melting points, may range from soft and flexible to quite rigid, and are strong electrical conductors.

Crystalline solids are three-dimensional collections of individual atoms, ions, or whole molecules that are structured in repeating patterns. These atoms, ions, or molecules are known as lattice points and are commonly represented as round spheres.

A solid's two-dimensional layers are formed by packing the lattice point "spheres" into square or closed packed arrays. By stacking the two-dimensional layers on top of each other, a three-dimensional lattice point arrangement represented by a unit cell is formed. A unit cell is the smallest group of lattice points that may be repeated to form a crystalline solid.

The solid can be imagined as the consequence of stacking several unit cells together. The kind of layer (square or close-packed), the manner each succeeding layer is put on the layer below, and the coordination number for each lattice point (the number of "spheres" contacting the "sphere" of interest) all influence the unit cell of a solid). Metals are shiny, opaque solids that are malleable and ductile. They are malleable if they are soft and can be molded or pressed into thin sheets, and ductile if they can be drawn into wires.

Valence electrons are not transferred or exchanged in a metallic link, as they are in ionic and covalent bonding. Rather, the electron clouds of neighboring atoms overlap, causing electrons to become delocalized. Throughout the crystal, electrons flow with relative freedom from one atom to the next.

A metal can be thought of as a lattice of positive cations surrounded by a "sea" of negative electrons. Metals are extremely conductive of heat and electricity due to their electron mobility.

Metals often have high melting points, with noteworthy exceptions including mercury, which has a melting point of minus 37.84 degrees Fahrenheit (minus 38.8 degrees Celsius), and phosphorous, which has a melting temperature of 111.2 degrees Fahrenheit (44 degrees Celsius).

A solid alloy is a blend of a metallic element and another material. Alloys are more workable than pure metals, which might be too malleable and hefty. Bronze is a copper and tin alloy, whereas steel is iron, carbon, and another additional alloy.

7.4.5. Amorphous Solids

The particles in amorphous solids (literally "solids without shape") do not even have a recurring lattice structure. These are sometimes known as "pseudo solids." Glass, rubber, gels, and so most polymers are examples of amorphous solids.

Figure 7.6. Amorphous solids.

Source: Image by PxHere

Because the bonds do not break all at once, an amorphous solid does not have a defined melting point; instead, it melts gradually across a range of temperatures. This indicates that an amorphous solid will melt into a soft, flexible state (similar to candle wax or molten glass) without entirely transforming into a liquid.

Because amorphous materials lack symmetry, they do not have regular planes of cleavage when cut; the edges may be bent. Isotropic materials have qualities such as refractive index, conductivity, and tensile strength that are the same regardless of the direction in which a force is exerted.

7.5. SOLID STATE OF CROSS-LINKED MACROMOLECULES

Cross-linked systems of flexible macromolecules have long been recognized as posing unusual challenges to statistical mechanics. These issues stem from a combination of two factors: the molecules' chainlike structure and the presence of persistent cross-links between the chains. These very same two mechanisms are considered to be responsible for the remarkable elastic responsiveness of cross-linked systems like gels and rubber.

Researchers argue that a strong knowledge of the undeformed solid state is required for a realistic statistical mechanical explanation of the elastic response of these systems. The purpose of research is to clarify the physical arguments that underpin the newly proposed theory of the liquid-to-solid transition in cross-linked macromolecules. Although much is known about gels and rubbers from a phenomenological standpoint?

A true microscopic theory must deal with the topological complexity of interconnected systems. The fundamental issue is clear. a section of a cross-linked system in which two chains, A and B, are cross-linked to the network so that chain A is in front of chain B.

The chains cannot travel through one another or the remainder of the system, the chains cannot move into the configuration indicated as the system experiences its dynamics. Chain B is in front of chain A in this configuration, yet the chains are cross-linked at the same places.

The two configurations are considered to be topologically inequivalent; if the system is established in the configuration, the topological connections between the chains will always be preserved. Because of the network of cross-links and the impenetrability of the chains, the system dynamics can never enable chain A to get ahead of chain B. They are intertwined.

That's the problem: how can one do statistical mechanical computations while keeping in mind that the network has just one topology? The goal of study is to describe the newly suggested solution to this problem in basic physical terms.

A number of efforts have been made to resolve the same question; most of them fall into one of three categories:

1. Disregarding the problem,
2. Tube models: as well as
3. Topological invariants

Approaches 1 and 2 have been demonstrated to be valuable phenomenological tools, but they are not claimed to be systematic theories. Approach 3 involves serious mathematical challenges, which will be addressed further below, and it has not been implemented beyond the basic model issues. The primary idea behind tube models is to confine the chain within a random tube to keep it near to some notional mean location in the lattice. The restricting effects of nearby chains are shown by the random tube.

The technique based on topological invariants posits that each network topology has a collection of invariants. A topological invariant is a number, 7, which can be calculated using a specific process for any configuration C of the network's chains. This technique will be as a function T. Thus, 7 = T.

The topological invariant is the same for all topologically similar configurations. Moreover, no two network topologies provide the same topological invariant value. Thus, knowing the value of the topological invariant allows one to uniquely determine the network's topology; similarly, knowing the network's topology allows one to uniquely assess the value of the topological invariant. The statistical mechanics of the system in a particular topology may then be determined while keeping the topology constant. The partition function, 2, is determined by the topology selected; consequently, 2 is a function of 7. Where kB is Boltzmann's constant, Ti is the temperature, and E(CJ) is the system's energy in configuration C. Unfortunately, no known techniques have all of the features. All of the previous "invariants," including the Gaussian invariant and the Alexander polynomials, have various topologically inequivalent configurations that produce the same numerical value for 7. As a result, these invariants cannot be applied to T. (C). Even if this were not the case, imposing the condition would very likely be impossible. Here, we will propose that the above-mentioned challenges may be addressed by considering the space, s, of all chain configurations in the system. We shall demonstrate this. Each system topology, compatible with a particular set of cross-links, creates a distinct subregion of S. This is equal to the prescription and is hence, in general, impossible to execute. Nevertheless, for many uses, the partition function contains more information than is necessary. For example, in this study, we will consider whether a system with a specific number of cross-links is a liquid or a solid. We will illustrate how to construct an order parameter with the following properties:

1. It can distinguish between the three possible thermodynamic phases, liquid, crystalline solid, and equilibrium amorphous solid; and
2. It can be expressed by summing a certain quantity over all configurations in S, rather than just those configurations within one specific subregion of S that correspond to one specific topology.

This order parameter is designed in such a manner that it exhibits a signature of how S is partitioned into subregions that correspond to the network's potential topologies.

Researchers were capable of approximating this order parameter as a function of crosslink density in gelatin or latex and discovered that when the cross-link density surpasses a threshold value, there is still a second-order transition phase from a liquid to an equilibrium amorphous solid. These computations have been briefly reported elsewhere and will be the topic of a comprehensive publication in the near future. The purpose of this section is to offer a physically justified description of this theory without obscuring technical specifics. As a result, we'll primarily stick to descriptive arguments. It should be noted, however, that the logic is exact and does not, in principle, depend on assumptions. It is especially true for realistic actual E((C) options. When one simply attempts to utilize the principles to compute the phase diagram for a system of randomly cross-linked macromolecules, one will be compelled to make approximations, such as using the Edwards Hamiltonian for E ((C)).

7.6. STRUCTURE OF CONFIGURATION SPACE FOR A CROSS-LINKED SYSTEM

Researchers now describe simple physical factors that influence the structure of configuration space in a system of cross-linked macromolecules. There are two major factors. The first is concerned with the manifestation of the numerous alternative topologies accessible to a system once the cross-links have been described in configuration space. The second batch of issues is connected to the possibility of phase transitions occurring in the system. They will be particularly interested in the transition from a liquid to a solid state.

Ergodicity is the primary premise of equilibrium statistical mechanics that we will employ. A statistical mechanical system is considered to be ergodic if, over an arbitrarily long time period, it examines, with arbitrary precision, every configuration possible to it.

An ideal gas with infinitesimally weak interactions in a container of finite volume V, for example, is ergodic. The particles may explore every location in the volume with any momentum during an arbitrarily long time period, subject only to the constraint that energy is conserved.

A system made of cross-linked macromolecules is not ergodic, as we will demonstrate below: In reality, ergodicity can be violated to variable degrees depending on the presence of topology and the system's phase.

7.6.1. TOPOLOGY

Considering a finite-volume V container filled with long flexible chains. Let us first assume that the chains are not cross-linked and do not interact. The system has access to all potential configurations, and it will denote the space of all possible configurations by S.

Assume that perhaps the system is randomly cross-linked. Cross-links permanently bind two monomers on two (or maybe the same) chains to the same point in space, however, this position is not set. Because of the limitation, many configurations in S are now unreachable.

For example, if a cross-link exists between chain 26 at monomer 135 and chain 14 at monomer 333, no configurations in which these two monomers do not occupy the same location in space are permitted. As a result, the system of cross-linked but noninteracting chains explores an area of configuration space, S, that is a subset of S. The system is ergodic, but only in a limited number of configurations.

Furthermore, imagine switching on the hard-core repulsions between the chains one at a time. If the reader is unhappy with this and may require that the cross-linked monomers be within one atomic distance of one another or that the chains be infinitesimally thin.

The crucial thing to remember is that the chains cannot cross through each other during their dynamics. Whenever the connections are turned on, the chains are in configuration C with some topology. Following that, the chains must be permanently trapped in that topology.

This suggests that the chains' available configurations are a subset of S. The collection of configurations available to the cross-linked, interacting chains will be referred to as Sz. The system is still ergodic, but it now extends across S2, which is a subset of S1. S2 is the set of all topologically comparable configurations to C.

C', the system may have been in a different configuration at the time. If C and C' are not topologically comparable, the system will be ergodic across the set of topologically equivalent configurations to C'.

This set might be called Si. In general, we can see that when we switch on the interaction, there will be many different subregions of configuration space in which the system might become stuck. S2 and SI2 are just two of an unlimited number of subregions. Szfi, p = 1... m will be used to represent them.

Because a particular configuration of the chains cannot be in more than one topology at the same time, these subregions of configuration space will be discontinuous. As a result, we can observe that the combination of cross-links and chain impenetrability has fragmented the original configuration space S into smaller, discontinuous portions such as S2 and S'2.

The system will be ergodic only across configurations confined in one of these subregions. To recap, the consequence of cross-links and impenetrability is that the system becomes imprisoned in a subregion of the configuration space of a cross-linked but penetrable set of chains.

7.6.2. Phase Transition

Let's really consider what happens to a statistical mechanical system during a phase transformation. Let's begin by studying a system of point particles; after the fundamental concepts have been established, we will immediately apply them to a system of cross-linked macromolecules.

Our focus is the phase transition to the solid state; thus, we'll need to describe what a solid is in statistical mechanics. It should be defined as a solid in two ways that are closely connected. The very first definition says that a solid can withstand minuscule static shear but a liquid cannot. In other words, if a solid is progressively sheared, it will apply a restoring force.

The importance of the phrase static is as follows: If one shears a liquid, there will be a restoring force initially, but if one waits long enough, the force will go away. A solid, on the other hand, will always have a restoring force.

The second meaning is thermodynamic and pertains solely to an infinite system in thermal equilibrium. This is because the solid form suddenly destroys the system's Hamiltonian translational invariance. Remember that the Hamiltonian H is equivalent to the sum of the system's kinetic energy K and potential energy V.

The potential energy is usually determined only by the difference in locations between the particles, rather than their absolute positions. As a result, shifting the coordinate origin has no effect on potential energy. Similarly, shifting the origin of coordinates has no effect on the kinetic energy, which depends on the time derivative of the particles' coordinates.

As a result, the Hamiltonian is translationally invariant. However, even if the Hamiltonian is translationally invariant, the system state does not have to be. The crystalline state is a well-known example. In contrast to a gas or liquid, the atoms are confined in a periodic pattern there.

In fact, even if the atoms are concentrated at randomly distributed points in space, translational invariance is still lost. In summary, the commencement of the solid phase happens when the system's state (as given by the density matrix, for example) spontaneously breaks the Hamiltonian's translational invariance. The stiffness of the solid state is an effect of this symmetry breakdown.

A solid is not rigid because it is held stiff by long-range forces. Rather, it is also the choice of the atoms to localize themselves in certain positions with regard to their neighbors in order to reduce free energy that gives stiffness to the solid.

Let's really look at how the liquid-to-solid transition appears in configuration space. Atoms in the solid state are concentrated around their mean location. Many combinations are possible in the liquid state that is not available in the solid state. The collection of settings accessible to the solid will be referred to as S3. S3 is a subdivision of s.

This, however, is not entirely right. Because the Hamiltonian is translationally invariant, if all the atoms are centralized around a location, say, 10 in., the system may solidify in a variety of ways. That arrangement would be as beneficial energetically if they were moved away from their mean locations given by So. In reality, after we've picked a certain system state that breaks, by merely translating the initial state, one may construct an endless number of alternative, equally acceptable ones. The same observations apply to rotations, as the Hamiltonian is generally rotationally symmetric as well.

Whenever a liquid-to-solid transfer takes place, the liquid's configuration space is divided in subsets Si where I = 1... m. The Hamiltonian symmetries connect the configurations in each subset S.

That example, if taken one configuration from one of the subsets, say Sand translate it, say, 10 in., In S399, scientists shall produce all setups. When scientists state that the subregions Si are connected by symmetry, they imply that executing the symmetry operation of the Hamiltonian to all the configurations in one of the subsets yields all the configurations in all the other subsets.

Now apply this image to the crystallization of a cross-linked macromolecule. Although shown, the existence of topology indicates that the system examines a subregion of the configuration space accessible to a cross-linked yet noninteracting collection of links. Let us imagine the system is ergodic in the Sp subregion.

Assume that by modifying a Hamiltonian parameter, such as the pH of the solvent, we may induce the system to solidify. The subregion of phase space S2 will then further fragment into a set of smaller subregions, S,'. Moreover, translational and rotational symmetry connects these subregions.

As a result, the initial configuration space of the cross-linked but noninteracting set of chains, S1, has been fractured into subregions S2', it has in turn been broken into subregions Si. Examine a specific subregion, say S237, from the collection of subregions S2'.

This is a collection of S3 type subregions that emerged from the spontaneous breaking of translational invariance inside the topology corresponding to S2l3I. As a result, this group of S3 subregions is linked by symmetry. Consider two S3 subregions, S31557 and S317860, for example; the former is contained with S2l3I, whereas the latter is contained with S226.

Because the topologies of these S3-type subregions differ (because they are contained within different S2-type subregions), they have no symmetry between them. Lastly, it has been noted that while our ergodicity-breaking strategy wasn't the most generic, it is definitely the simplest that one can imagine.

This ergodic method is, in fact, the suggested explanation of the computation. There is no reason to presume a priori that all subregions of the S2 type located within a particular subregion are connected by symmetry; the computation.

7.7. CONSTRUCTION OF AN ORDER PARAMETER

The configuration space for a collection of cross-linked impermeable macromolecules may grow rather intricate, especially when the solid state is reached. When performing statistical mechanical averages, therefore, only those configurations across which the system is genuinely ergodic must be included.

This necessitates understanding how to specify configurations in each subregion of configuration space. It is unknown how to achieve this in a sophisticated system like the one under discussion here; nevertheless, it is doable in smaller systems like Ising ferromagnets.

In this part, the authors explain another order parameter that may be calculated using mean-field theory by taking a statistical mechanical average of all the configurations in S1 and which determines whether S1 is divided into subregions S2 and S3. The approach was invented by Parisi, who used it to solve the infinite-range issue. Using a spinning glass.

7.8. PHYSICAL STATES AND MOTIONS OF SMALL MOLECULES

When considering molecular mobility, it is important to remember that molecules do not exist in isolation from one another. A molecule's mobility affects other molecules that are somewhat distant from it. In other words, scientists progress from understanding individual molecule motion to examining the motion of molecules within a molecular body.

The body, whether supramolecular, molecular, atomic, ionic or metal in structure, is a material object of a higher order than the particles of which it is made, according to all three motion criteria:

1. The body is made up of molecules; it evolved from molecules; independent of individual motion of molecules, they all (more or less) cooperate in the motion of the body to which they belong.
2. The space required for body motion is bigger than the volume filled by individual molecules.

3. The qualitative, quantitative, and genetic relationship between repulsion and attraction between bodies has been elevated to a higher degree.

In physics parlance, there exist "gravitational forces. "Long-range and low-strength attracting forces exist between bodies, whereas intermolecular attractive and repulsive forces of short-range and high intensity exist between molecules.

The mobility of molecules inside a body is less than the motion of the body as a whole. Nevertheless, chemistry and polymer science do not address the mobility of the entire body (it is a subject of mechanics). As a result, in this chapter, we will solely look at the mutual mobility of molecules within the body.

When discussing the physical (i.e. aggregate) condition of matter, one cannot speak exclusively about individual molecules and their mutual motion. It is not possible to say that one molecule is solid while another is liquid since the aggregate state is the product of all the molecules' interactions.

The motion of an individual molecule in a solid is the outcome of the mutual motion of that molecule and surrounding molecules, which is simplified to the interactions of attraction and repulsion in its most generic form. As a result, each aggregate state is differentiated by a different ratio of attraction to repulsion.

Attraction and repulsion can refer to attractive and repulsive forces (as defined by Newton and Boscovich) or the distance between particles, i.e. Density or thinning of matter (Hegel's understanding), or approaching and distancing (Engels' understanding).

Particles (molecules or atoms) are aggregated to an extremely dense state in a solid substance. Because the solid state has the greatest inter-atomic and/or inter-molecular interactions /3/, translational motion is blocked. At very low temperatures, the ratio of intermolecular repulsion to attraction is totally switched to the side of attraction as a result of energy loss (cooling).

That is why the single molecules do not move as a whole. As the temperature rises, molecules or groups of atoms begin to oscillate about some equilibrium locations, i.e., molecules approach and move further away from one another.

The amplitudes of the oscillations are tiny, but the frequency is high: 1013 - 1014 oscillations per second /4/. The molecules get more energy as the solid body heats, and the amplitudes of the oscillations increase. They rise so much near the melting point that the molecules contact, i.e. the space required for molecular motion equals the available space.

After the body melts, a liquid state is formed in which the molecules are also tightly packed (attracted), because the density is only lowered by a few percent /4/ during the transition from the solid to the liquid state.

There are revolving aggregations of molecules in liquids (so-called clusters). Such short-lived aggregations emerge and depart all the time, so as the temperature rises, they get smaller. These clusters breakdown around the boiling point into single revolving molecules.

Particle aggregation is insignificant in the gaseous state at low pressure. In comparison to their diameters, the distances between molecules are quite enormous. Molecules move quickly in translation, rotation, and oscillation /4/. Intermolecular attraction forces are minimal in gases.

Nothing appears to hinder single molecules' unfettered translational motion. The attraction appears to have been totally vanquished, and the repulsion has triumphed - the molecules extending, repelling, and separating themselves until the whole available volume is occupied.

However, this is an incorrect conclusion. The attraction cannot be eradicated, nor can the repulsion be removed from the attraction altogether. In general, dialectics rejects the idea that opposites can be totally separated or that one may completely dominate the other, so that the other suffers divided or that one might entirely dominate the other, causing the other to vanish.

So, what exactly is intermolecular attraction in gases? What restricts the movement of molecules in a gas? It's the molecules colliding with each other! (At 20°C and atmospheric pressure, each molecule in a gas collides with other molecules a billion times each second.

It happens right in front of our eyes, but we don't see it: in the air, molecules of nitrogen clash with molecules of oxygen, and molecules of nitrogen smash with molecules of oxygen. All hydrogen molecules collide 1, 91029 times per second in 1 cm^3 of gas /5/.) These collisions are responsible for limiting the mobility of each individual molecule.

Thus, attraction is not eliminated in gases, but rather converted into a qualitatively and quantitatively distinct form than attraction in liquids. The oneness and fight of these two opposites: free mobility of molecules (repulsion) and random collision of molecules (attraction) are used to explain many phenomena (chemical reaction rate, viscosity, heat transfer, and diffusion).

Collisions in gases cause a wide range of lower-level molecular movements, including the rotation of the molecule as a whole, the internal rotation of the atoms in the molecule, the vibration of atoms bonded in the molecules, electron excitation, and so on. These are all slower modes of motion than the overall translation of molecules.

Thus, by heating the solid from absolute zero, the oscillatory motion of the molecules happens first, followed by rotational motion in liquids and lastly translational motion in gases. The general norm is that a particle that absorbs motion does so in its most basic form first, and then progressively progresses to higher forms.

By cooling the gas, the molecules initially lose their translational motion, condense into a liquid in which they may spin, and then further cooling generates a solid state in which the molecules still can oscillate. At absolute zero, all molecular motion ceases. The general rule is that a particle that loses motion loses it first in its greatest form, and then progressively in all of its lesser forms.

The body is an item composed of smaller particles that work together to move the body as a whole, and by motion, we mean the body's approximation and distance from another body. This type of motion is known as mechanical motion.

It refers to the exterior movement of the body. However, the body's component particles are also moving in an unruly manner. This internal motion is responsible for the development of new body attributes (pressure, temperature, density, and so on) that cannot be traced to individual molecules. Internal particle motion in the body is a lower sort of motion than exterior body motion.

Those two types of motion are diametrically opposed: Mechanical motion achieves the maximum cooperative motion in a solid body because all molecules (atoms, ions) move in the direction of the body motion (ignoring oscillations that are irrelevant to the motion of the body).

The gaseous body has the least cooperative motion because one component of it might travel in one direction and the other in the opposite direction at the same time. Following the collision, the two solid entities will decompose, with just a small percentage of the mechanical motion being converted to molecular motion. The collision of two gaseous substances, on the other hand, will result in their full mixing. Given Aristotle's belief that one is more one if its motion is more one /1/, the more cooperative the movements of its components, the higher the form of motion of a whole. As a result, solid body motion is of greater order than liquid or gaseous body motion. However, from the perspective of the molecules in the body, the situation is exactly the reverse.

Molecules have the highest motion in a gaseous state, but molecules have essentially little motion in a solid body. The movement of molecules inside a gas is a higher kind of motion in terms of the molecules themselves, with a higher amount of repulsion and attraction than the motion of molecules in liquids or solid bodies.

7.9. PHYSICAL STATES AND MOTIONS OF MACROMOLECULES

A substance's transition to a gaseous form is done by evaporation, boiling, or sublimation. Individual molecules detach from the surrounding molecules and transition from the liquid or solid to the gaseous phase.

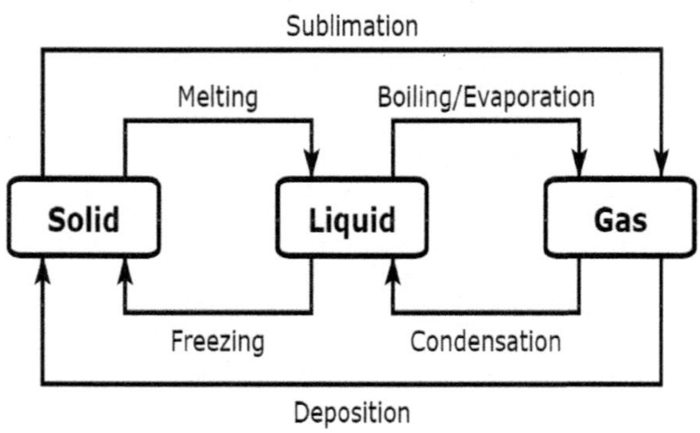

Figure 7.7. Diagram describing matter phase transition.

Source: Image by Wikimedia Commons

Because the macromolecular chains are exceedingly entangled and have multiple intermolecular forces related to other chains or other molecules (e.g. solvents), it is hard for the individual chains to escape these clamps and pass into the gaseous phase. By introducing the thermal energy required to transform the polymer melt into a gaseous state, no evaporation but mostly polymer breakdown happens. (As a result, there is no free translation of single isolated macromolecules in their whole.)

There are three sorts of solid states for macromolecules: crystalline (i.e. partly crystalline), glassy, and rubber-like. The kind, speed, and energy of the motion of the macromolecule and its components differ between these states. Polymers are capable of the following forms of motion:

a. The translational motion produced by the flow of the entire macromolecule;
b. Rotation and jumps of segments (parts) of a macromolecular chain with a length of 40-50 carbon atoms;
c. Rotation of several atoms in the main chain and of short lateral branches, end groups, or substituents;
d. Vibration (oscillation) of segments of macromolecules or atoms around equilibrium positions.

At low temperatures, the energy of the macromolecules is insufficient to allow rotational and translational motions of the segments, therefore only vibrational motion (oscillations) occurs, resulting in changes in atomic distances and valence angle distortion.

As the temperature rises, the peaks of the vibrational motion rise, and eventually the thermal energy of the macromolecules approaches the energy barrier for the rotational and translational motion of segments with lengths of 40–50 carbon atoms, and the glassy polymer transforms into a rubber-like (viscoelastic) state that exists only in macromolecular substances.

Some chain repeating units (i.e. residual monomer), atomic groups, and longer segments of macromolecules are vulnerable to substantial thermal motion in the rubberlike state, but macromolecules as a whole, as independent kinetic units, are doubtful. In the case of linear macromolecules, it is assumed that the presence of entanglements prevents the translational motion of complete macromolecules.

These entanglements, in our opinion, are a type of attraction of repeating units of chains, as a modified version of collisions, i.e. attraction that the monomer molecules possessed prior to being included into the chain. The existence of chemical connections between the chains of cross-linked macromolecules prevents them from moving.

As the temperature rises over the flow temperature, the linear rubberlike polymer will transition to a liquid state, resulting in intense thermal motion of the repeating units and large portions of the polymer chain as well as the displacement of whole macromolecules as a result of mechanical force-induced coordinated motion of its segments, the crosslinked polymer behaves differently from linear polymers.

Heating does not contribute to the shift from glassy to liquid due to the presence of crosslinks, which are chemical interactions between macromolecular chains that prevent complete macromolecules from translating. The crystalline polymer enters a liquid state above the melting temperature of the crystalline phase and acts like any other non-crosslinked polymer.

Heating solid glassy or crystalline polymer bodies, similar to low-molecular-weight substances, produces oscillatory motion of small parts of macromolecules, then rotational motion of segments in the rubberlike state, and finally translational motion of whole macromolecules in the polymer's liquid state.

The general rule is likewise satisfied in polymers, i.e. the particle that receives the motion initially acquires it in its most basic form, and then gradually progresses to higher forms.

On the opposite, cooling the melt causes the first loss of translational motion of the entire macromolecules, resulting in a rubberlike state in which the segments of the chains can rotate, and further cooling results in a solid glassy or crystalline state in which parts of the macromolecular chain can only oscillate.

Any movement of the macromolecular chains ceases at very low temperatures. The general norm is that a particle that loses motion loses it first in its greatest form, then progressively in lesser forms.

7.10. CONCLUSION

In the case of periodic elements, the valence bond continuity can be the source of unique electronic properties, a subject of most recent topicality. When the chains are rigid, a unique liquid crystalline state may result.

In this latter context, the distinction is being made between linear chain molecules and networks, between flexible and rigid molecules and finally between macromolecules that have a periodic molecular constitution and those which have not.

The consequences of factors a-f are illustrated in the lecture by specific examples. In the lecture, this is followed by a broad classification of polymeric matter. This is the intrinsic reason for macromolecules being the principal constituents of living organisms containing and transmitting genetic information and regulating metabolic processes.

These are chosen so as to relate to problems and methods in various branches of physics ranging from mechanics to superconductivity, from statistical mechanics to biophysics, and from particle physics to crystal growth. First, it is emphasized, that, although the specialty area polymer science is a meeting point virtually of all-natural sciences as regards the field of relevance and the methodology used in its practice.

Macromolecular substances respond in a rather unique fashion to radiation due to the fact that owing to the large molecular size a single event within one molecule affects a large amount of material.

In the case of the random state of the molecule and in the case of chain flexibility the enormous variety of conformational possibilities gives rise to a unique state of matter characterized by exceptionally high elastic extensibility, this elasticity being entropic in character rubber elasticity.

REFERENCES

1. Bagley, M., 2014. *Properties of Matter: Solids*. [online] livescience.com. Available at: <https://www.livescience.com/46946-solids.html> [Accessed 30 June 2022].
2. Chem.fsu.edu. n.d. *Properties of solids*. [online] Available at: <https://www.chem.fsu.edu/chemlab/chm1046course/solids.html> [Accessed 30 June 2022].
3. Goldbart, P., 2017. Issue Editorial Masthead. *Macromolecules*, [online] 50(5). Available at: <https://guava.physics.uiuc.edu/~nigel/REPRINTS/1989/Goldbart%20Solid%20State%20of%20Cross-Linked%20Macromolecules%20Macromol%201989%20%28PDF%29.pdf>.
4. Jetybayeva, A., Uzakbaiuly, B., Mukanova, A., Nurpeissova, A. and Bakenov, Z., 2022. Solid-State Nanobatteries. *ACS Symposium Series*, [online] pp.201-248. Available at: <https://pubs.acs.org/doi/abs/10.1021/bk-2022-1414.ch010> [Accessed 30 June 2022].
5. Keller, A., 1983. Polymer Physics: Organization of Macromolecules in the Solid State. *Physik Journal*, [online] 39(7), pp.178-180. Available at: <https://onlinelibrary.wiley.com/doi/pdf/10.1002/phbl.19830390706> [Accessed 30 June 2022].
6. Ladizhansky, V., 2018. Solid-State NMR of Macromolecules. *Reference Module in Chemistry, Molecular Sciences and Chemical Engineering*, [online] Available at: <https://www.sciencedirect.com/science/article/pii/B9780124095472140831?via%3Dihub> [Accessed 30 June 2022].
7. Stoiljkovic, D., 2022. *Attraction And Repulsion In Polymer Science. Part I. Attraction and repulsion in macromolecules formation, according to common interpretation*. [online] Academia.edu. Available at: <https://www.academia.edu/44945585/ATTRACTION_AND_REPULSION_IN_POLYMER_SCIENCE_Part_I_Attraction_and_repulsion_in_macromolecules_formation_according_to_common_interpretation> [Accessed 30 June 2022].

CHAPTER 8

Advances in High-Temperature Network Polymers of Carboranylenesiloxanes and Silarylene-Siloxanes

CONTENTS

Researchers began researching the impact of incorporating closodicarbaborane units into chains and networks of nearly every known form of polymer in the 1960s and 1970s. In polymeric systems, carborane was shown to generate significant thermal stability in the product. Since Kipping and colleagues discovered siloxanes, classical polysiloxanes – [(R)(R') SiO] n –, or silicones, have been intensively investigated and some of them marketed as early as the 1940s.

The electronegativities of silicon and oxygen are 1.8 and 3.5, respectively, as calculated by Pauling. This difference is maximum in the Si–O bond – Si–O– having an estimated 37 to 51 percent ionic character. It is a partly ionic bond with some features of a double bond. These two fundamental properties of the siloxane bond are essentially responsible for its high heat stability and unique chemical activity.

In polymeric systems, carborane was shown to generate significant thermal stability in the product. The Hydrosilylation reaction is the insertion of Si–H bonds to organic unsaturations such as olefins, acetylenes, and ketones.

8.1. INTRODUCTION

Since Kipping and colleagues discovered siloxanes, classical polysiloxanes – [(R)(R') SiO]n –, or silicones, have been intensively investigated and some of them marketed as early as the 1940s. Their wide set of qualities has permitted their use in industries as diverse as aeronautics, biomedical, cosmetic, surface water-proofing, sealant, and unmolding agents.

Silicones' most noteworthy qualities are their extremely low glass transition temperatures (Tg), which derive from the flexibility of their backbones, and their low surface tension. These two characteristics are responsible for their extensive variety of applications.

Silicones also have high heat stability, which can be attributed to their essential structural feature, the siloxane bond –Si–O–. It is a partly ionic bond with some double bond properties. The former attribute derives from the unusually substantial difference in electronegativities among silicon and oxygen, which Pauling estimated to be 1.8 & 3.5, respectively.

This difference is high in an approximate 37 to 51 percent ionic nature of the Si–O bond. The latter, on the other hand, is associated with partial overlap of both the vacant low-energy silicon d orbitals with oxygen p orbitals, which occurs due to the relatively large size difference between these two atoms, allowing oxygen to back-donate its lone two electrons and form a dπ-pπ bond in addition to the normal σ bond.

These two focal characteristics of the siloxane bond are, essentially, the reasons for its pronounced thermal stability and specific chemical behavior Nevertheless, under unambiguous conditions like in a destructive or a base medium, or at high temperature, a silicone backbone containing polarizable siloxyl units is exposed to degradation by ionic reactions.

In view of such reactions, the backbone is seen to depolymerize by chain scission of a piece of the - Si-O-Si moieties through a six-centered framework achieving the improvement of the thermodynamically steadier low-sub-nuclear weight cyclic things and more restricted direct chains (Figure 8.1).

This intramolecular cycloreversion or depolymerization of silicones has been observed to take place from as few as four consecutive Si–O bonds. To alleviate this issue of depolymerization, many groups of hybrid silicones were synthesized as early as the 1950s, and were generated through inclusion of large and bulky groups such as alkyl, aryl, alkyl, aryl, carboranyl, and fluoroalkyl in the backbone of silicone precursors. It was theorized that the presence of such large rigid groups would provide pronounced resistance to depolymerization by preventing formation of cyclic siloxane volatile products, a result of making the reaction considerably more difficult.

Among the hybrid silicones that were made, the most comprehensively researched were the Silarylene-siloxanes and the Carboranylenesiloxanes (Figure 8.1). In each of these silicone groups a percentage of oxygen atoms in the parent backbone are substituted with aromatic or carboranyl groups, respectively

Because of carboranes, a couple of investigation packs during the 1960s began looking at the effect that the combination of closodicarbaborane units into chains and associations of basically any realized sort of polymers would have.

It was soon recognized that the carborane units attached as pendant groups to the otherwise organic or inorganic backbone of the polymer produced no practical advantage. However, if the carborane became part of the polymeric chain (e.g., regularly interspersed among repeating –Si(R)2 Ounits) they induced an appreciable thermal stability in the product . In polymeric systems, it was observed that the meta and para congeners of the closodicarbaborane

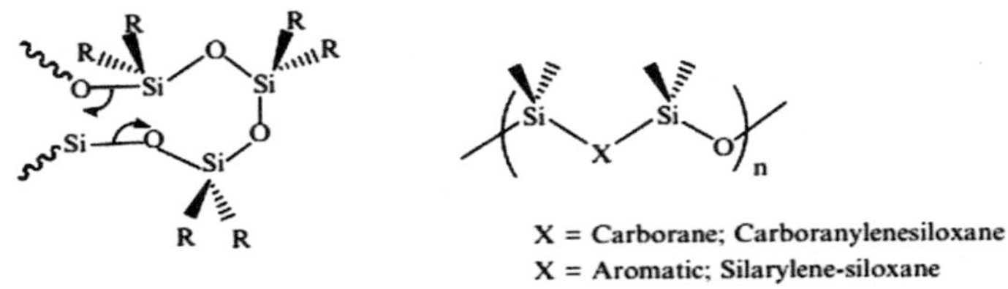

Figure 8.1. Schematic representation of a six-centered mechanism for depolymerization of siloxanes and hybrid siloxanes.

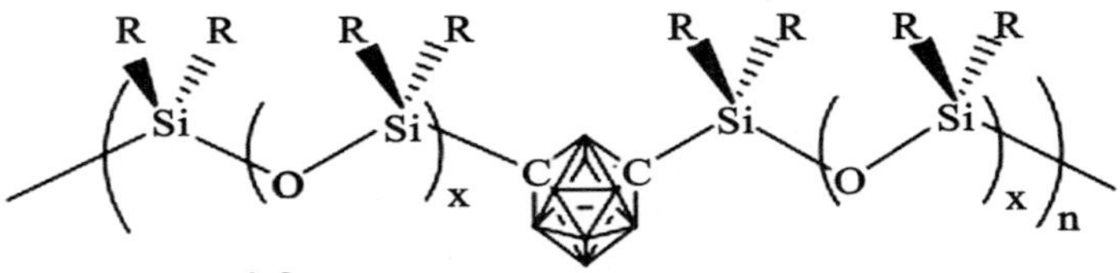

Figure 8.2. DEXSIL polymer.

With the formula C2 B10H12 conferred the best properties comparative with other carborane atoms, the ortho congener was seen to shape stable monomeric, or, best case scenario, dimeric, cyclic species that were of no utilization in polymer development. The best group of thermally stable high-atomic weight half and half Carboranylenesiloxanes polymers were gotten from the group of poly-m-Carboranylenesiloxanes, which became eminent under the business trademark DEXSIL (Figure 8.2).

Around a similar time, a few groups endeavored to integrate aromatic units into natural and inorganic polymers. Similarly, as with carboranes, it was seen that there was an obvious improvement in the warm soundness of silicones just when the aromatic groups were integrated into the backbone and not as pendant groups on the silicones.

The joining of aromatic groups into the backbones of silicones was accomplished through buildup reactions of disilanol monomers with different receptive silane monomers. Four head techniques were created: (1) the chlorosilane course, (2) the aminosilane course, (3) the acetoxysilane course, and (4) the ureidosilane course, in view of the receptive usefulness on the siloxane monomer that responded with the liquor usefulness on the siloxane reactant to yield the cross-breed silicone.

The improved performance of these two groups of hybrid silicones was the direct consequence of the carborane or aromatic-imparted strengthening of the Si–O bonds and their described impedance of the depolymerization reaction. Additionally, in the case of the carboranylenesiloxanes, the thermo-oxidative stability of the boroncontaining carboranes also contributed to the polymer's improved stability. However, it was soon recognized that forming thermally robust siloxane polymers through inclusion of large and rigid molecules between the flexible siloxane segments resulted in overall reductions in polymer elasticity.

Thus, even in its beginning phases, innovative work of cross breed silicones was designed for getting ready designs which exploited the high-temperature adjustment while at the same time keeping up with the low temperature properties of polysiloxanes. The age of arranged or vulcanized forms of the created half and half silicones was sought after to deliver polymers of further developed mechanical and warm properties that likewise had upgraded processability.

The vulcanization of polysiloxanes was generally accomplished through one or the other an "actuated fix," in which different crosslinking specialists were used along with strain and intensity, or through the "room-temperature vulcanization" (RTV) processes in which no outer energy was fundamental.

The majority of the enacted fix strategies were known to utilize free radicle initiators, for example, natural peroxides to cause crosslinking reactions. The RTV fixes were accomplished either by a buildup reaction in which receptive polymer end groups, for example, silanol or acetoxy silane units were joined in the presence of of a catalyst, or by platinum-catalyzed Hydrosilylation reactions of pendant groups like vinyl or allyl.

Comparative courses were likewise investigated for the vulcanization of carboranylenesiloxanes and Silarylene-siloxanes. Most of the reports on the vulcanization of carboranylenesiloxanes polymers and the properties of their vulcanizates were worried about the m-CB10H10C-containing polymers. A few different unsaturated groups were inspected as expected receptive destinations for the vulcanization of

Carboranylenesiloxanes polymers. The revealed groups included natural unsaturation like vinyl, allyl, and vinyl-o-carboranyl units. It was found that the best blend of properties was gotten, especially before heat maturing, from the vulcanization of the vinyl-o-carboranyl groups. Be that as it may, this approach was subsequently deserted as a result of the troubles in planning such mixtures.

Vulcanization of carboranylenesiloxanes was additionally endeavored through reactions with natural peroxides at high temperatures and tensions. One model is the formation of an organization polymer through polymerization of the vinyl groups of a pendant vinyl-containing Carboranylenesiloxanes by natural peroxides in air at 315 °C for 300 hours. This polymerization could likewise be advanced under tension, bringing about finishing in around 2 hours.

Room temperature vulcanization of m-CB10H10C Carboranylenesiloxanes elastomers has likewise been endeavored and detailed in view of by the same token: (1) platinum-catalyzed silylhydride-vinylsilane reaction or (2) the reaction of tris (acetoxy) silane with carboranylsilanol ended prepolymers. Notwithstanding, neither of these strategies was created to the degree of viable use.

The vulcanization of silphenylene-siloxane polymers was accounted for at room temperature either in arrangement or in mass with silicate esters. For crosslinking in arrangement, toluene was utilized as dissolvable, and the added substances utilized for restoring comprised of 20 pieces of to some extent hydrolyzed ethyl silicate and 2.5 pieces of dibutyltin diacetate to 100 pieces of the polymer.

In another report, the crosslinking of silarylenesiloxane polymers, which contained p-phenylene and p′-diphenyl ether units in their Silarylene groups, and different measures of vinyl siloxane groups with pendant vinyl groups were performed utilizing dicumyl peroxide as the crosslinking reagent in light of its low reactivity to methyl side groups and its high reactivity toward vinyl groups.

These polymers were intended to allow controlled vulcanization for the readiness of elastomers with explicit crosslinking densities to empower an examination of their mechanical properties as elements of various crosslink densities. A report containing two vulcanization reactions of vinyl containing Silarylene-siloxane polymers utilizing in one case 2,5-dimethyl-2,5-di (t-butyl-peroxy) hexane and in the other dicumyl peroxide as the crosslinking specialist is likewise accessible.

In these vulcanizations, intensified gum-stocks were first processed and afterward vulcanized under undefined tension for 20 minutes at 170 °C, trailed by one more pattern of 16 hours at 232 °C.

As of late, notwithstanding, examination into mixture silicones of Silarylene-siloxanes and carboranylenesiloxanes has been centered around improvement of half breeds with cross linkable groups in the foundation of the polymer. Keller and colleagues changed the science of the Carboranylenesiloxane framework to integrate unsaturation inside the backbone, or at the terminal locales, as a hotspot for the polymerization to a thermoset or network.

The created forerunners contained either cross linkable diacetylene groups (in oligomers) or RTV natural groups like vinyl or ethynyl bunch (in monomers) which empower their transformation into broadened network polymers. The organization polymer for each situation was produced by the warm polymerization of the diacetylene bunch or by the Hydrosilylation reaction of vinyl or ethynyl units, individually.

The underlying instances of the thermally-and hydrosilatively-framed Carboranylenesiloxane and Silarylene-siloxane network polymers were seen to have plastic attributes. Later changes in the engineered techniques for formation of such materials have brought about the advancement of elastomeric variants of organization polymers for both of the frameworks.

8.2. THEORY: SOLID-STATE POLYMERIZATION OF DIACETYLENE GROUPS

Despite the fact that the polymerization of diacetylenes in the half breed silicones depicted in this section fundamentally isn't led in their strong states, a comprehension of the strong state polymerization of diacetylenes is considered helpful for a more prominent enthusiasm for the crosslinking processes happening in these mixture silicones.

The diacetylene strong state polymerization has been concentrated widely by Wegner. The polymerization is known to deliver polydiacetylenes with a broad p-formation in the backbone. The accessible methods of polymerization for diacetylenes are the 1,2-expansion, 1,4-expansion, and cyclic trimerization reactions.

From his examinations during the 1970s, Wegner reasoned that the most predominant component in the diacetylene strong state polymerization is the 1,4-expansion system (Figure 8.3). The polymerization is normally done by exposing the monomer crystals to heat, bright (UV) radiation, or by high-energy illumination.

A detailed description of diacetylene solid-state polymerization theory is beyond the scope of this chapter. In general, the theory includes a least-motion criterion which predicts the solid-state reactivity of different diacetylene phases and the most likely polymerization mode for a particular phase from a two-parameter description of the relationship between neighboring monomer molecules. The theory also considers the phase stability criteria for characteristic free energy diagrams that explain the phase behaviors that have been observed

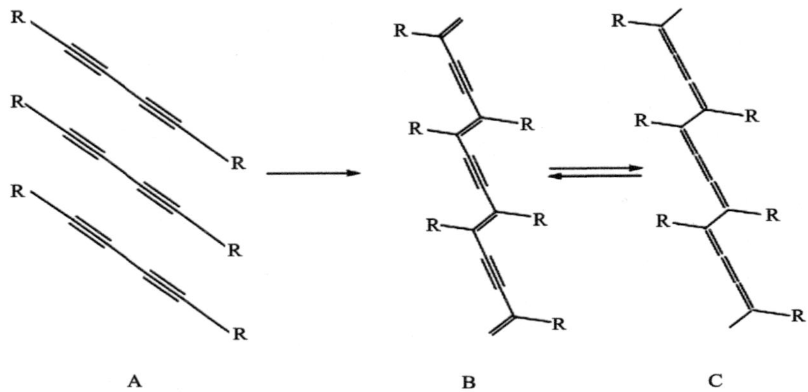

Figure 8.3. Topochemical solid-state polymerization of diacetylenes (1,4-addition). Parallel diacetylene molecules (A) reacting to form the trans, trans polymer, which can be represented by the alternate monomeric structures (B) and (C).for different polymerizing diacetylenes. It additionally considers the reaction uniqueness rules as they connect with monomer site point bunch balance, evenness relations between commonly responding monomer particles, and layered changes during polymerization. It has been seen that during a diacetylene strong state polymerization reaction, as the reaction continues, a homogeneous arrangement of the polymer chains is framed in the strong monomer framework. The polymers are typically splendidly shaded and are for the most part insoluble like natural solvents in which the monomers are dissolvable. The distinction in the level of dissolvability of the monomer and the polymer in like manner solvents has been used to extricate the unreacted monomer from the to some degree polymerized tests and to get a proportion of the monomer to polymer transformation.

8.3. THEORY: HYDROSILYLATION REACTION

The expansion of Si-H bonds to natural unsaturation like olefins, acetylenes, and ketones is known as the Hydrosilylation reaction. A greater part of the

Hydrosilylation reactions is catalyzed by soluble transition metal structures and are known as homogeneous Hydrosilylation reactions.

Synergist Hydrosilylation reactions are known to be exceptionally intricate reactions. Be that as it may, a couple of speculations have been drawn about the mechanism of such reactions (Figure 8.4). Homogeneous olefin Hydrosilylation is expected to begin with a required oxidative expansion of the Si-H bond in a cis style to the reactant metal. This cycle is followed by a hydride transitory inclusion that is trailed by an irreversible reductive end of the end result. The oxidative expansion and transitory addition steps are known to be reversible in nature. Heterogeneously catalyzed Hydrosilylation reactions are likewise known.

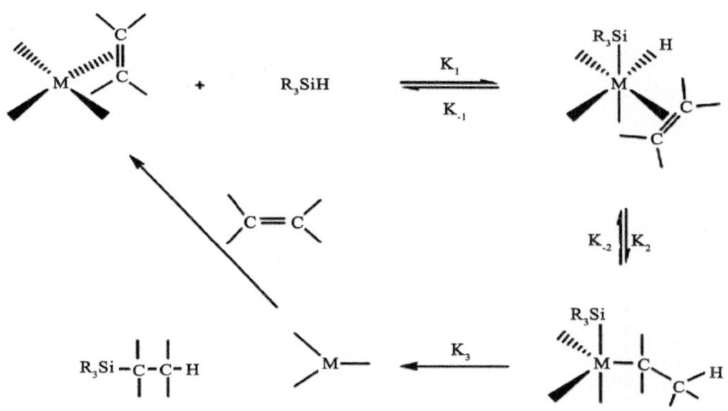

Figure 8.4. A generalized mechanism for catalytic Hydrosilylation of olefins.

Heterogeneous catalysts, for example, the Speier's catalyst and the Karstedt catalyst have been known to work by comparative heterogeneous systems wherein the development of fine chemically dynamic colloidal Pt particles is accepted to happen during the catalyst commencement step. Hydrosilylation reactions are significant in the production of silicones.

Hydrosilylation, otherwise called hydrosilation, is perhaps the most valuable synergist reaction prompting the development of organ silanes and organosilicons, which have an assortment of utilizations in industry and as intermediates in natural science.

Hydrosilylation happens through the expansion of H-Si to an unsaturated bond, for example, carbon bond, carbon-oxygen bond, carbon-nitrogen bond, nitrogen bond and nitrogen-oxygen bond utilizing a metal catalyst ion, Lewis Acid, or a catalyst.

8.4. THEORY: CARBORANES

Carboranes are compounds of boron, carbon, and hydrogen in which both boron and carbon atoms are incorporated into three-dimensional, polyhedral skeletons of the general formula $C_pB_qH_{p+q}$. All known individuals from this family could be arranged into three significant groups: the closo-, nido-, and arachno-carboranes, where the first address the C0−2BqHq+2 intensifies which are worked of closed polyhedral enclosures, while the last two incorporate those carboranes in which the boron-carbon polyhedra look like the states of nest skeletons.

The closed shell designs of the closo-dicarbaboranes are known to add to their astounding synthetic dormancy, particularly to acids. Accordingly, icosahedral carboranes are not gone after by hot sulfuric acid and are basically inert to the customary combination of nitric and sulfuric acid that easily nitrates most aromatic species.

Clearly the steadiest closo-dicarbaboranes are the three individuals from the icosahedral family, the 1,2-, 1,7-, and 1,12-dicarba-closo-dodecaboranes better known by the more famous names o-, m-, and p-carborane, separately.

A weak point of carboranes is their susceptibility to alkaline degradation. This shortcoming applies for all intents and purposes just to the species with neighboring carbon particles. In this manner, o-carborane is quantitatively changed over completely to the nido-[C2B9H12] − particle inside a couple of hours by sodium methoxide in ethanol or by sodium (or potassium) hydroxide in ethanol.

With m-carborane a similar kind of debasement continues no less than multiple times all the more leisurely and a temperature of no less than 70 °C is expected for any down to earth impact; the primary noticed corruption of the para isomer must be constrained involving a 30% arrangement of potassium hydroxide in propylene glycol at 180 °C.

Carboranes, by and large, and icosahedral carboranes, specifically, in their impartial and anionic structures are likewise referred to for their excellent qualities like low nucleophilicity, high hydrophobicity, and their electron-pulling out properties having profoundly polarizable σ-aromatic groups.

The carboranes have polyhedral sub-atomic designs in light of organizations of boron and carbon atoms, in which the carbon particles possess neighboring positions.

Accordingly, the designs of carboranes and their subordinates are like those of the isoelectronic (having similar number of electrons) polyhedral boranes, and, similar to the boranes, they include three-focus bonds as well as conventional two-focus bonds. Their most huge primary element is the covalent holding of carbon all the while to five or six different atoms.

8.5. CARBORANYLENESILOXANE POLYMERS CONTAINING THERMALLY CROSSLINKABLE OR VULCANIZABLE DIACETYLENE GROUPS

Since Wegner's investigations on strong state polymerization of diacetylenes, organosilicon polymers with spanning diacetylene units were being orchestrated and detailed by 1990. The interest in these polymers containing unsaturated natural units spanned by substituted silicon molecules fundamentally originated from their conductivity and nonlinear properties.

The underlying announced models comprised of just silane or silylene polymers, and not siloxane polymers. The principal illustration of the joining of a diacetylene bunch into a siloxane polymer was accounted for in 1994 by Henderson and Keller.

For the incorporation of the diacetylene group, the dilithiadiacetylide formation route utilized by Barton et al was used to generate the reactive lithiated unsaturated organic unit which was subsequently reacted by a condensation reaction with a halogenated carboranylenesiloxane to produce the poly (carborane-siloxane-acetylene) polymers. In the synthesis which was a one-pot, two step reaction, hexachlorobutadiene was cleanly converted to dilithiobutadiyne through reaction with 4 equiv of n-BuLi in tetrahydrofuran (THF) and was subsequently reacted with an equi molar amount of the precursor carboranylenesiloxane, the Dexsil monomer, 1,7-bis(chlorotetramethyldisiloxy)-m-carborane, to generate the linear polymer 1 in high yield (Figure 8.5).

The siloxane unit in the carboranylenesiloxane was the disiloxyl unit. The GPC examination of the polymer 1 which is dissolvable in most normal natural solvents had shown the presence of low-atomic weight species (Mw = 400) as well as higher sub-atomic weight polymers (Mw = 4,900, Mn = 2,400) averaging to around 10 recurrent units.

Warming of 1 at 150 °C under decreased pressure brought about the expulsion of lower sub-atomic weight polymeric species, leaving the item in a 92 to 95% generally speaking yield. The FTIR range of 1 (Figure 8.6; top) displayed unmistakable retentions at 2,963 (C-H), 2,600 (B-H), 2,175

(C≡C), 1,260 (C-Si), and 1,081 (Si-O) cm−1. The absorption of the internal diacetylenes at 2,175 cm^{-1} was found to be as intense as the other

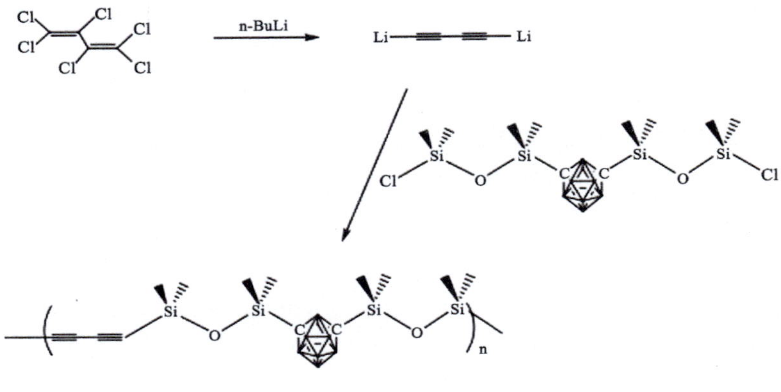

Figure 8.5. Synthesis of the poly(m-carborane-disiloxane-diacetylene) (1) reported by Henderson et al.

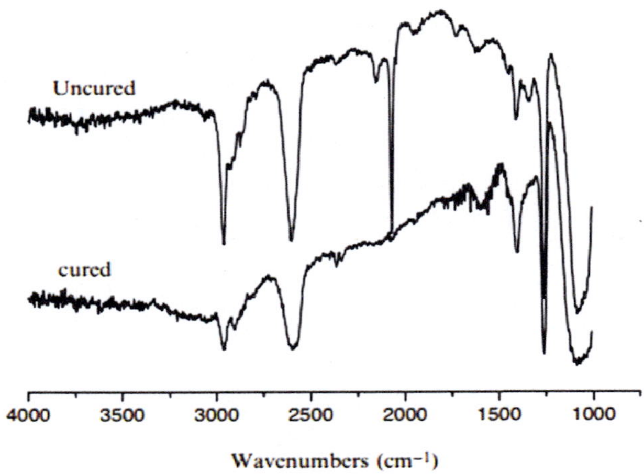

Figure 8.6. FT-IR spectrum of 1 (top) and the network produced from 1 (bottom).

Vibrational assimilations of 1 despite the fact that inward alkynes are frequently known to have exceptionally powerless or nonexistent advances.

Under heat circumstances, the liquid linear polymer 1 was found to easily transform into a thermoplastic by the bridging of diacetylene units, as indicated by the elimination of the acetylenic absorption at 2,175 cm1 and the emergence of a new, moderate peak centred at 1,600 (C=C) cm1, indicating a 1,4-addition process (Figure 8.6; bottom). The DSC study found that diacetylene interconnecting began at 250 °C and increased at 341 °C (Figure 8.11a). The totally crosslinked product was reported to be plastic in character and to lack a glass transition temperature.

The crosslinked material was pyrolyzed in gas at 1,000 °C to obtain a black solid ceramic material with an 85% yield. Despite this, pyrolysis of the crosslinked product in gas at 1,000 °C yielded a ceramic with a 92% efficiency, which was most likely attributable to weight gain from oxidising the borons on the outer layer to B2O3, resulting in a protective outermost surface. The ceramic products were evaluated using SEM, X-ray photoelectron spectroscopy, Raman, and scanning auger microscopy scatter tests. In comparison to material oxidised at 500 °C, specimens heated inertly to 400 °C and 900 °C showed negligible or reduced surface inorganic segregation. Carbon dominated the surface and mass of specimens burned in gas to 400 °C, containing trace amounts of oxygen, boron, and silicon The microstructures and outer layer formed by 500 °C oxidation and responsible for the char's resistance to subsequent high-temperature corrosion were determined to be composed of an antioxidant property boron oxide and silicon oxide bilayer formed after 500 °C oxidation. The SiO2 layer was intended to act as an oxygen barrier, whilst boron oxide was considered to be capable of filling fissures caused by thermal expansion incompatibility with the underlying bulk. Long-term oxidative exposure at 500 °C caused boron oxide evaporation, resulting in a SiO2 layer on the outer layer.

Son et al. reported the high-yield production of comparable diacetylene polymers containing just siloxane groups in 1995, shortly after Henderson et al. The polymers' disiloxane 2a and trisiloxane 2b versions were produced using the simple dilithiadiacetylide production method. The maximum yield of the synthesis represented an improvement over the synthesis of the disiloxane-diacetylene polymer described by Parnell et al., which was generated by the chemical interactions of 1,3-diethynyltetramethyldisiloxane and produced mainly low molecular weight polymers as well as some insoluble material. Both polymers' GPC analyses revealed broad molecular weight ranges, with peak maxima at 10,000 (relative to polystyrene). This research is essential because it clearly illustrates the efficiency of the dilithiadiacetylide production technique in integrating diacetylene units into siloxane polymers. Moreover, it provides a standard for evaluating diacetylene polymers of carboranylenesiloxanes.

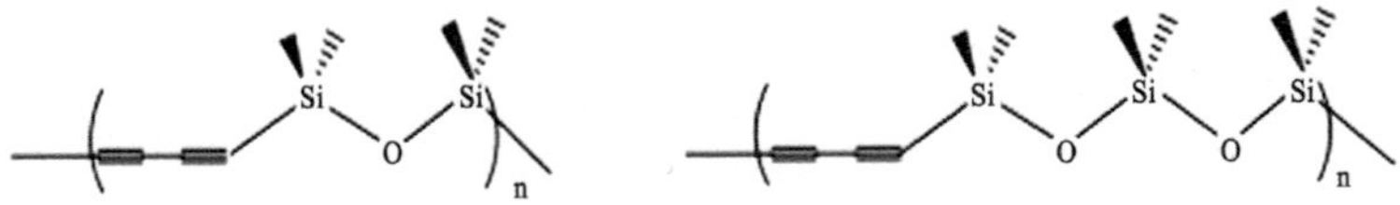

2a 2b

Figure 8.7. The poly(diacetylene-disiloxane) and poly(diacetylene-trisiloxane) systems of Son et al.

in comparison to their parent siloxane versions, DSC analysis revealed that the disiloxane and trisiloxane polymers were vulcanised by the heat curing responses of their diacetylene groups, with exotherms reaching a maximum at 289 and 315 °C for the disiloxane and trisiloxane polymers, respectively, to produce tough, void-free thermoplastics with Tg values ranging from 144 to 170 °C.

In comparison, zero glass transitions were seen in the thermoplastic generated during the curing of diacetylene-containing carboranylenesiloxane polymers over 450 °C.

Sundar et al. adopted the dilithiadiacetylide production technique later in their published synthesis of linear boron-silicon-diacetylene copolymers in which they investigated another boron source, namely, phenyl boron dichloride (PBD), for carborane in the synthesis of diacetylene polymers of boron-containing siloxanes (Figure 8.8).

Figure 8.8 summarises the compositions of the boron-silicon-diacetylene copolymers 3a-d. The mole per cent ratios of silane/siloxane (DMS, TMDS, and HMTS) to boron (PBD) were reported as 90:10 and 80:20, respectively. They discovered that when alternative siloxanes to boron ratios (such as 40:60 and 50:50) were utilised in the synthesis, mainly low-molecular-weight compounds were produced.

The separated copolymers 3a-c were found to be solids at room temperature, while 3d, possessing the trisiloxane group, was stated to be a thick liquid.

Since the copolymers 3a-c were also not totally soluble in THF, the molecule weight was determined using polystyrene as a reference just on the trisiloxane copolymer 3d and was reported to be in the 2,300 g/mol range. Char yields of 77.4, 75.0, 72.1, and 44.4% were achieved after thermal processing of the four polymers to 1,000 °C in air. DMS (3a: 77.4 wt%) > TMDS (3b and 3c: 75.0 and 72.1 wt%) > HMTS (3d: 44.4 wt%)

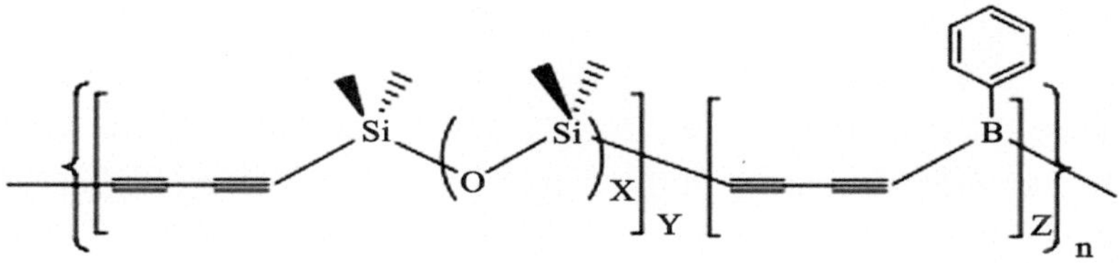

	X	Y	Z
3a	0	0.8	0.2
3b	1	0.8	0.2
3c	1	0.9	0.1
3d	2	0.9	0.1

Figure 8.8. The general structure of the (boron-silicon-diacetylene) copolymers of Sundar et al. (top) and the compositions of the representative polymers (bottom).

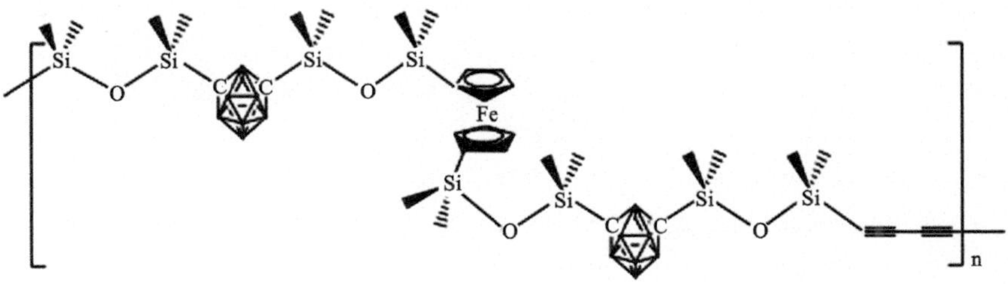

Figure 8.9. The linear (ferrocenyl-carboranylenesiloxyl-diacetylene) polymers of Houser et al.

All of the polymers formed thermoplastic after interconnecting reactions of their diacetylene units in an anaerobic environment at 400 °C. Yet, no mention of the thermosets' origin was provided. According to the findings of a DSC investigation, the exotherms for the crosslinking of diacetylenes in 3a-d ranged from 200 to 225 °C, with a peak maximum ranging from 285 to 300 °C.

In 1998, Houser et al. reported the creation of diacetylene-containing carboranylenesiloxanes with certain carborane groups substituted by linking organometallic ferrocenyl groups 4 (Figure 8.9). The polymers were synthesised by reacting dilithiobutadiyne with two equivalents of 1,7-bis(chlorotetramethyldisiloxyl)-m-carborane (Dexsil monomer), then treated using dilithioferrocene-tmeda (1 equiv). According to GPC studies, this substance has a molecular weight of around 10,000 compared to polystyrene, which accounts for the existence of ~10 repeat units in the polymer. Thermal treatment at 350 °C in an inert environment produced a black, elastomeric thermoset with 98% weight retention. In the product's DSC thermogram, the curing of the diacetylenes showed an exotherm rising at 280 °C. A modest percentage of shrink was noted during the thermoset's production. The thermoset's elastomeric character contrasted sharply with the stiff and robust thermosets made from siloxyldiacetylene polymers like poly (tetramethyldisiloxyl-diacetylene) published by Son et al. or poly(carborane-tetramethyldisiloxyl-diacetylene) reported by Henderson et al. As a result, this was the first example of a diacetylene-containing siloxane system that formed an elastomeric networked thermoset. Increase the temperature of the thermoplastic materials by 1,000 °C. Further heating of the thermoset under N2 to 1,000 °C resulted in a solid, black ceramic with a 78% ceramic output.

In search of elastomeric networks of diacetylene-containing carboranylenesiloxane polymers, Kolel-Veetil et al. investigated the effects of diacetylene unit concentration dispersion in the parental poly(carborane-disiloxane-diacetylene) on the plasticities of the associated networks in 2003. The concentration/ratio of the carborane, disiloxane, and diacetylene moieties in the parent polymer's repeating unit was 1:2:1.

The network polymer/thermoset derived from this polymer has plastic qualities at room temperature and also no visible glass transition temperature in the product. The research was founded on the concept that the plasticity of the parent system's curing networks was due to the increased density of crosslinkable diacetylenic groups in the parent polymer that had generated a substantially crosslinked network upon curing, resulting in a considerably restricted flexibility of the-Si-O-Si-backbone flexibility. It is thus hypothesised that decreasing the proportion of diacetylenic groups in the backbone of the precursor linear polymer 1 will result in an elastomeric networked material following curing.

As a result, the elastomeric characteristics of networks generated from linear poly(carborane-siloxane-acetylene) must be adjustable by altering the concentration of the diacetylene unit in the precursor polymer. The report covered the fabrication and thermal characterisation of three linear hybrid poly(carborane-disiloxane-diacetylene) systems, 5a-c, with different carborane: siloxane: acetylene ratios (5a [2:3:1], 5b [4:5:1], and 5c [9:10:1]. A series of blocky and alternate polymers with the same carborane:siloxane:

acetylene ratios were produced (Figure 8.10). Two oligomeric stages, one a lithium end-capped carboranylsiloxane intermediary and the other chlorine capped diacetylenesiloxane intermediary, were synthesised and united during the production of each blocky polymer to produce the finished sticky brown poly(carborane-disiloxane-diacetylene) in high yield (85-95%). (Figure 8.10; right) Comparable alternate polymers were created by reacting lithiated diacetylene with the appropriate dichlorodisiloxane-capped m-carborane (Figure 8.10; left).

Both approaches resulted in the formation of linear polymers. The *in-situ* creation of a desirable carboranylenesiloxane oligomer through the interaction of a lithiated carborane and a chlorinated siloxane molecule, instead of the use of the commercial Dexsil polymer, was a distinguishing characteristic of both syntheses.

This has allowed for a tremendous amount of variety in the formulations of diacetylene-containing carboranylenesiloxane monomers that Henderson et al. did not have.

The influence of the diacetylene concentration decrease was seen in the locations and intensities of the DSC peak maxima of the 5a-c crosslinking endotherms. Even as the percentage of diacetylene units in the polymer fell from 5a to 5c.

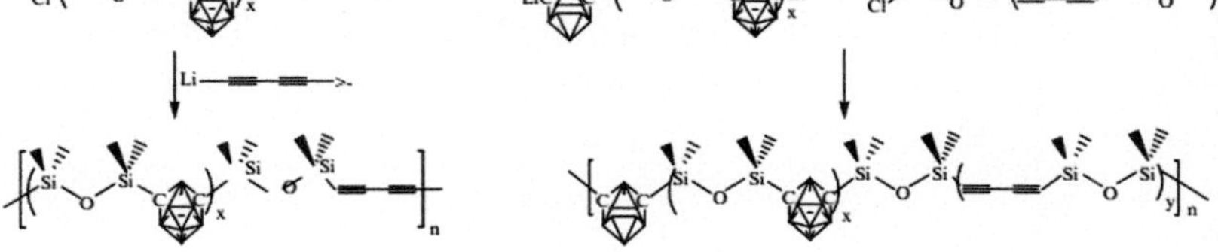

Figure 8.10. The synthetic schemes for the diacetylene-diluted (a) alternating (left) and (b) blocky poly (m-carborane-disiloxane-diacetylene)s reported by Kolel-Veetil et al.

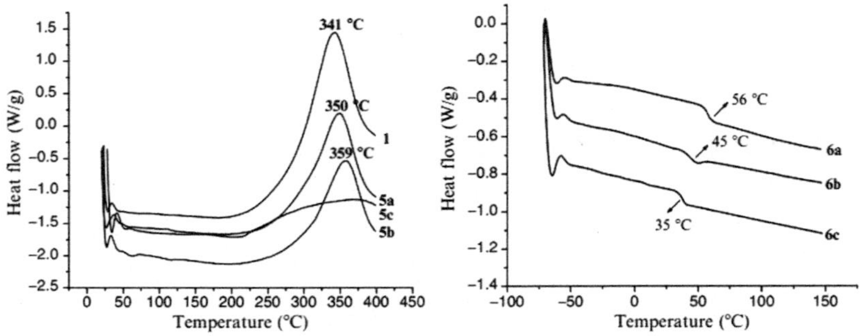

Figure 8.11. DSC thermograms of (a) the alternating poly(carborane-disiloxane-diacetylene) 5a (3:2:1), 5b (5:4:1), 5c (10:9:1), and 1(2:1:1) (left) and (b) DCS thermograms of glass transitions of the crosslinked networks 6a, 6b, and 6c produced from 5a, 5b, and 5c, respectively (right).

the peaked maxima of the related endotherm changed to a higher temperature and reduced in intensity (Figure 8.11; left), suggesting the need for more heat as well as a prolonged time to cure the diacetylene units in the diacetylene-diluted systems. As predicted, the high point maximum and intensity of the DSC exotherm of the parent poly(carborane-diloxane-diacetylene)s, 1, emerge at the lowest temperature and as the most intense exotherm in a sequence of comparison DSC exotherms of 1 and 5a-c (Figure 8.11; left).Tg values of the crosslinked networks 6a-c generated from the monomers 5a-c were found to be 56 °C, 45 °C, and 35 °C, respectively (Figure 8.11; right). Despite the fact that the Tg values of these networks decreased correspondingly with increasing diacetylene concentration in the parent polymer, these were found to be primarily plastic in character at room temperature. In comparison, it has been claimed that the covalently bonded network produced from 1 does not contain any glass transitions. This study demonstrated that the number of crosslinking groups in carboranylenesiloxanes can be used to control the flexibility of poly(carboranedisiloxane-diacetylene) networks.

Based on their findings with diacetylene-diluted carboranylenesiloxanes including trisiloxyl groups, Kolel-Veetil et al. produced and published comparable polycarboranylenesiloxane copolymers having trisiloxyl groups in 2004. The siloxane moiety was substituted in order to reap the benefits of a more flexible siloxane (trisiloxyl vs. disiloxyl) on the flexibility of the produced networks. Researchers reported the development of a series of alternated and block copolymers with the identical trisiloxane:carborane:d iacetylene molar ratios of 10:9:1, 5:4:1, and 3:2:1, , by using two synthetic techniques outlined in Figure 8.10.After heat curing, the 1,4-addition processes of the diacetylene units produced networks in both groups of copolymers. Despite the fact that the two processes produced materials with equal molecular weights and overall compositions, changes in the copolymer sequencing were revealed to induce considerable variances in the calorimetric glass transitions for the copolymers with the highest concentrations of the diacetylene group (i.e., the 3:2:1 and 5:4:1 copolymers) (Table 8.1). For a given reactant ratio, the alternating copolymers had slightly larger molecular weights than that of the block copolymers.

DSC thermograms for both groups of polymers were used to measure the glass transitions (Tg) of the cured networks (Figure 8.12). Both sets of polymers showed significant glass transitions below 0 °C (Table 8). Minor transitions were also observed in the alternated copolymers at 13 and 19 °C, as well as at 46 °C, and in the block copolymers at 60 and 65 °C. Tg was found to grow proportionately when the diacetylene content in a set rose, resulting in an increase in crosslink density. Tg values were found to be well characterised for the two copolymers with the minimum concentration of diacetylene groups (7A10:9 and 8B10:9). In fact, the glass transition temperatures (46 °C for 7A10:9 and 49 °C for 8B10:9) were identical to the Tg of 50 °C observed for the uncrosslinked polycarboranylenesiloxane, which contained m-carborane and hexamethyltrisiloxane units. The glass transition is kept rather well-defined in the alternated copolymers with greater diacetylene concentrations (7A5:4 and 7A3:2). The block copolymers with larger diacetylene content, on the other hand (8B5:4 and 8B3:2).

Table 8.1. GPC molecular weights and thermal properties of the cured alternating ($7A_{3:2}$, $7A_{5:4}$ and $7A_{10:9}$) and block ($8B_{3:2}$, $8B_{5:4}$ and $8B_{10:9}$) polymers

Copolymer	M_n(Kg/mol)	M_w(kg/mol)	T_g(°C)	Char yield (%)
$7A_{3:2}$	4.3	6.6	–30	70
$7A_{5:4}$	3.3	5.4	–39	65
$7A_{10:9}$	5.2	8.6	–46	46
$8B_{3:2}$	3.4	6.2	–27	74
$8B_{5:4}$	2.9	4.3	–34	70
$8B_{10:9}$	4.1	6.7	–49	52

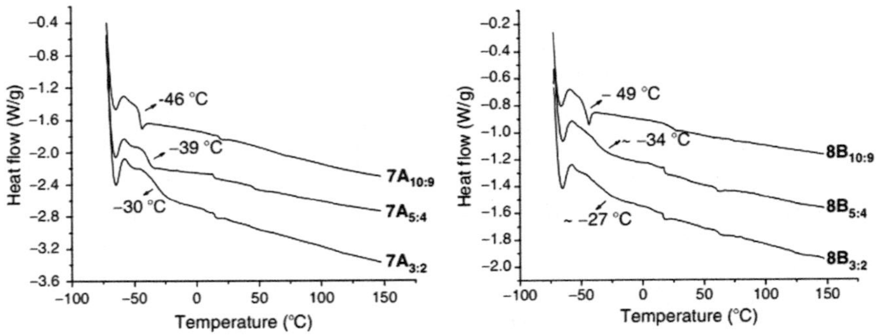

Figure 8.12. DSC thermograms of the glass transitions of the networks formed from (a) alternating and (b) blocky poly(m-carborane-trisiloxane-diacetylene).

When compared to samples with sharper glass transitions, the samples with broader glass transitions had a higher variety in segment lengths between cross connections. The thermal structural properties of the crosslinked networks were measured gravimetrically by heating to 1,000 °C in N2 (Table 8.1). Char yields for a given siloxane:carborane:diacetylene ratio were approximately 4 to 6% higher for block copolymers than for alternate copolymers. Therefore, our research proves that diacetylene-dilution and the replacement of disiloxane moieties by trisiloxane moieties in the parent diacetylene-containing carboranylsiloxane result in the formation of polymers which cure to form an elastomer network.

8.6. SILARYLENE-SILOXANE POLYMERS CONTAINING THERMALLY CROSSLINKABLE OR VULCANIZABLE DIACETYLENE GROUPS

Sundar et al. in 1997 presented the first samples of silarylene-siloxane polymers having crosslinkable diacetylene groups. (Figure 8.13). They described the polycondensation reaction of 1,4-dilithiobutadiyne with 1,4-bis(dimethylchlorosilyl) benzene and/or 1,7-bis(tetramethylchlorodisiloxane)-m-carborane, which resulted in a series of inorganic-organic linear diacetylenic hybrid polymers (9a-e). These polymers are actually a cross between carboranylenesiloxane and silarylene-siloxane. At room temperature, the polymers were thick liquids or low-melting solids but were soluble in ordinary organic solvents. DSC detected broad significant exotherms related to diacetylenic unit reaction in the 306 °C to 354 °C ambient temperature range.The covalent bonding of the diacetylene units was demonstrated to happen via a 1,4-addition process, similar to the scenario of the diacetylene-containing carboranylsiloxane systems. The thermoplastic nature of the networks formed by each of the monomers was discovered. The systems' molecular weights were not disclosed. Weight retentions of 79 to 86% were found when 9a-e were evaluated by TGA at 1,000 °C under nitrogen.

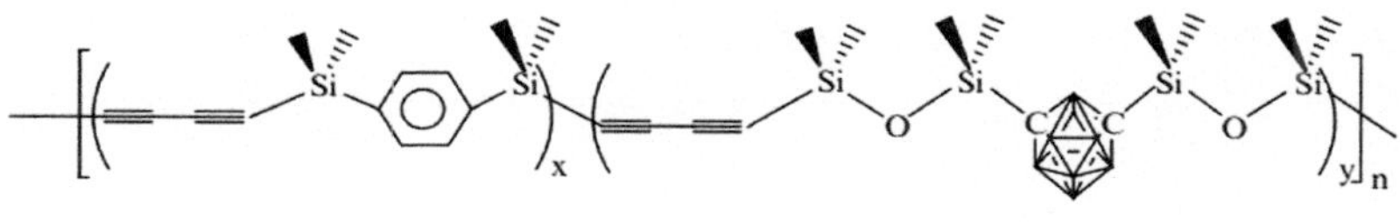

9a-e

a: x = 1, y = 0; **b**: x = 3, y = 1; **c**: x = 1, y = 1; **d**: x = 1, y = 3; **e** = x = 0, y = 1

Figure 8.13. The hybrid silarylene-siloxane/carboranylenesiloxane reported by Sundar et al.

The monomers 9a-e had charred outputs of 83, 86, 80, 73, and 71% when heated to 1,000 °C, correspondingly. The high charred production figures indicated the aliphatic unit's heat stabilising properties in the polymeric backbone. When heated to 1,000 °C in circulation of air, the briquettes formed from 9b-e were found to have no weight loss and even a minor weight increase (1-4%) resulting from the oxidation of boron to B2O3 and silicon to SiO2.

Homrighausen et al. published a study in 2002 describing the production of three categories of silarylenesiloxane polymers with diacetylene bridging groups. The initial paper described the creation and characterisation of a silarylene-siloxane-diacetylene polymer which, when cured, transforms into a thermosetting polymer. Polycondensation of 1,4-bis (dimethylaminodimethylsilyl)butadiyne with 1,4-bis(hydroxydimethylsilyl) benzene yielded the linear polymer 10. The approach was an adaptation of the aminosilane-deficient technique. (Figure 8.14).

It was discovered that throughout the production of linear polymer 10, some polymeric breakage happened at the alkynyl carbon-silicon link through dimethylamine reaction. This breakage process was shown to disturb the alternate character of the polymer structure, preventing the production of a really high-molecular-weight polymer. The average molecular weight of 10 was calculated to be around 10,000 g/mol. Transformation to a thermoplastic material was seen to proceed at 300 °C via diacetylene group bonding processes, with the exotherm reaching 336 °C in its DCS thermogram. On assessment, no glass transitions were identified in the cross-linked polymer.

The second batch of silarylene-disiloxane-diacetylene polymers 11a-c (Figure 8.15) published by Homrighausen et al. comprises a series of linear silarylene-siloxane-diacetylene polymers with changing diacetylene content created using a modified version of the aminosilane-deficient technique. The linear polymers were created by combining 1,4-bis(dimethylamino dimethylsilyl)butadiyne [(CH3)2 N-Si(CH3)2-C-C-C-(CH3)2 Si-N(CH3)2] with a sequence of disilanol prepolymers. Solution condensation of an excess of 1,4-bis(hydroxydimethylsilyl) benzene with bis(dimethylamino) dimethylsilane produced the hydroxyl-terminated low-molecular-weight silarylene-siloxane prepolymers. During production, the length of the prepolymer was adjusted by changing the molar ratio of 1,4-bis(hydroxydimethylsilyl)benzene to bis(dimethylamino) dimethylsilane. The spacing between the diacetylene units in the linear polymer was changed by varying the prepolymer length. As a result, the concentration of crosslinking in the thermally produced elastomers could be controlled.

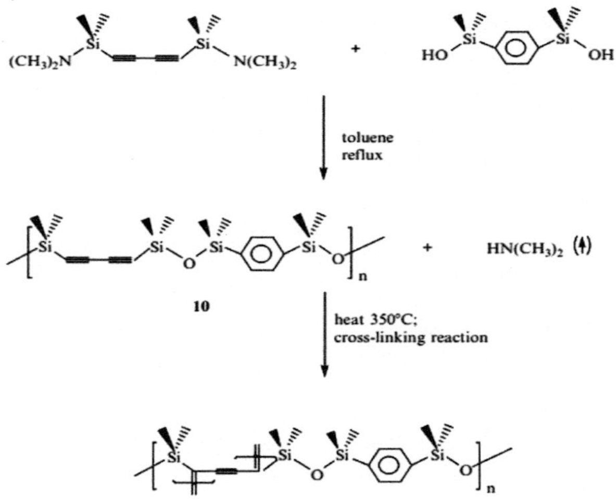

Figure 8.14. Synthesis of the thermosetting silarylene-siloxane-diacetylene polymer reported by Homrighausen et al.

The average molecular weight of 11b was measured by size exclusion chromatography to be 10,000 g/mol. Heat treatment of silarylene-siloxane-diacetylene linear polymers 11a-c resulted in a reaction between diacetylene units, resulting in covalently bonded (networked) polymers. The polymers which were linked were elastomeric compounds which were fully resistant to commonly used organic solvents. The researchers found that thermal treatment of polymers 11a-c to temperatures near or above the crosslinking reaction temperature resulted in the creation of soft, elastic, and void-free elastomeric materials. Therefore, the second set of diacetylene-containing silarylene-siloxane polymers reported by Homrighausen et al. proved the usefulness of reducing the concentration of diacetylene units as a technique of injecting elasticity into an otherwise plastic network.

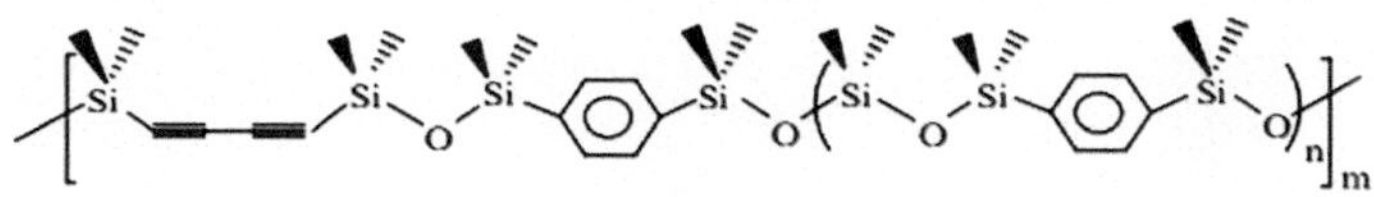

11a-c

11a n=1; **11b** n=2; **11c** n=3

Figure 8.15. Structure of the elastomeric diacetylene-diluted silarylene-siloxane-diacetylene polymer reported by Homrighausen et al.

Homrighausen et al. discovered a third set of silarylene-siloxane-diacetylene polymers that cured to form elastomeric networks (Figure 8.16). The parent polymers were created through a series of oligomeric, hydroxy-terminated silary lene-siloxane prepolymers and 1,4-bis(dimethylaminodim ethylsilyl)butadiyne [(CH3)2 N-Si (CH3)2-C] poly condensing processes.

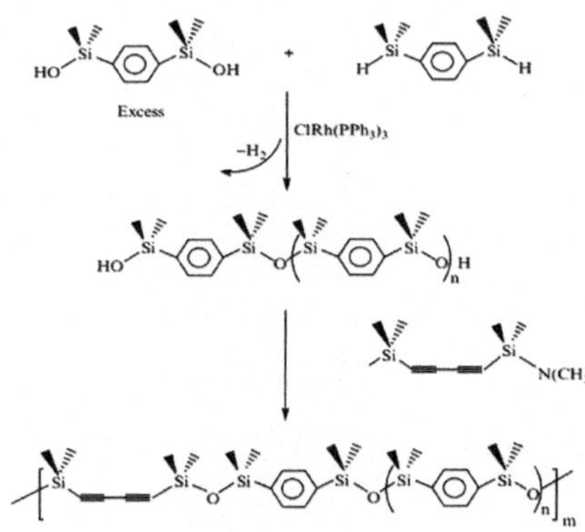

12a n=2, **12b** n = 4, **12c** n = 6, **12d** n = 8

Figure 8.16. Structure of the elastomeric diacetylene-diluted silarylene-siloxane-diacetylene polymer reported by Homrighausen et al. wherein Wilkinson's catalyst was used.

The oligomer was produced in the presence of a catalytic quantity of Wilkinson's catalyst [(Ph3P)3 RhC1]. Using a polystyrene reference, 12a-d had a number-average molecular mass varying from 10,000 to 15,000 g/mol. Thermal treatment of the linear polymers 12a-d of silarylene-diloxane-diacetylene proceeded in an interaction between diacetylene units, resulting in the covalently bonded (networked) polymers. In ordinary organic solvents, the connected polymers were observed to be fully resistant. Kolel-Veetil et al. revealed that by altering the length of the hydroxy-terminated prepolymer in the synthesis of 12, researchers were able to change the bonding strength. As a result, polymeric materials with varying degrees of crosslinking (elastomeric to plastic) might be created. The glass-transition temperatures for the covalently bonded polymers produced from 12a-d were determined by DSC to be 29°C, 22°C, 29°C, and 29°C, accordingly.

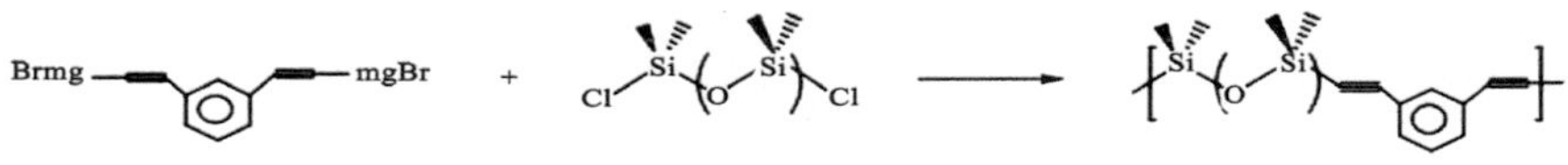

13

Figure 8.17. Reported synthesis of a poly (siloxylene-ethylene-phenylene-ethylene).

Grignard reagents have lately been utilised to generate hybrid silicones including phenyl and unsaturated organic groups, although not always as diacetylene groups. As per Figure 8.17, Poly (siloxylene-ethylene-phenyleneethylene)s (13) has been claimed to be synthesised by reacting a bischlorosiloxane with the bismagnesium derivatives of a diethynyl chemical as per the following method. These compounds have been found to be effective for composite materials with high heat tolerance.

8.7. HYBRID SILOXANE NETWORK POLYMERS FROM HYDROSILYLATION REACTIONS OF SILOXANE AND CARBORANYLENESILOXANE MONOMERS

Hydrosilylation reactions are a significant, if not the main, class of reactions used in the production of silicones. Such reactions may likewise be imagined in the amalgamation of elastomeric carboranylenesiloxanes and silarylenesiloxanes.

In 1998, Houser et al. announced the blends of the monomeric carboranylenesiloxane 1,7-bis(vinyltetramethyldisiloxyl)- m-carborane 14, and its Hydrosilylation reactions with the polymeric crosslinker, poly(methylhydrosiloxane) 15 (Figure 8.18). The reactions were catalyzed by the Speier's catalyst, H2 PtCl6. How much the crosslinker was fluctuated concerning how much the monomer to decide the proportion of the reactants that conferred the most noteworthy warm soundness to the item polymer. This was finished on the reason that the warm bond of a polymeric material relied mostly upon its crosslinking thickness. Three examples of 16 were ready with monomer to polymer proportions of 18.2, 9.3, and 4.8. The qualities related to vinyl to Si-H group proportions of 1.04, 0.53, and 0.27. The items were all observed to be hard and drab materials. None of the proportions was seen to yield an elastomeric item. Despite the fact that the reactions were performed with reliable reagents, the developments of the items were accounted for to require a day to a few days at room temperature. In light of a legitimate concern for creating elastomeric network polymers, Kolel-Veetil et al. detailed the adjustment of the previous Hydrosilylation reaction framework.

They revealed the Karstedt catalyst-catalyzed encompassing condition Hydrosilylation reactions of a monomeric vinyl 14 or ethynyl-containing 17 carboranylenesiloxane with three distinct monomeric expanded siloxane crosslinkers in hexane (Figure 8.19). The reactions including the vinylcarboranylenesiloxane were accounted for to produce a group of totally hydrosilated network polymers 18. On account of the ethynyl monomer, the reactions were done at two distinct proportions yielding a somewhat and a totally immersed set of organization polymers.

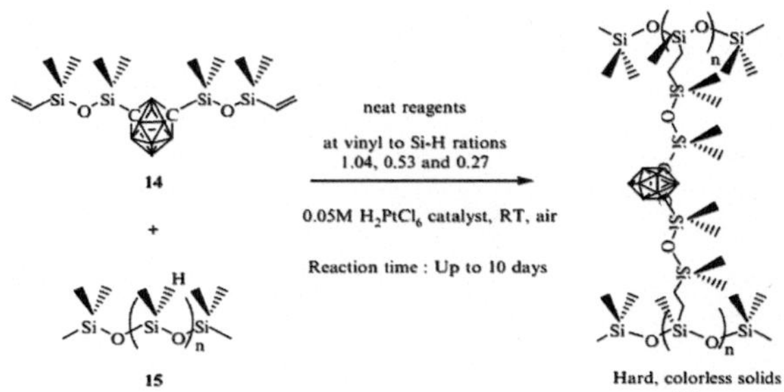

Figure 8.18. Hydrosilylation reactions producing hard, colorless network plastics reported by Houser et al.

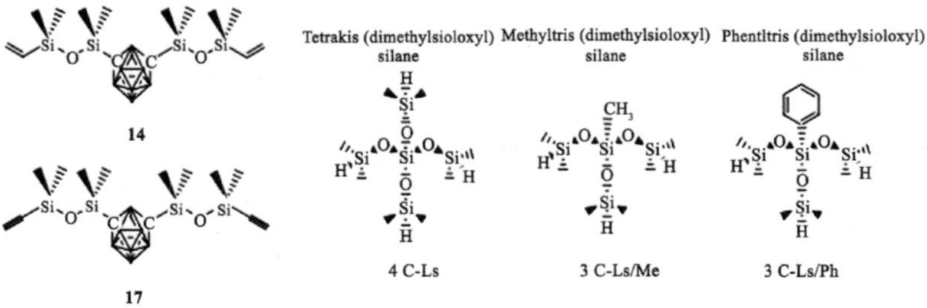

Figure 8.19. The carboranylenesiloxane monomers and the branched siloxane crosslinkers used in the Hydrosilylation reactions reported by Kolel-Veetil et al.

The rationale behind their decision of the Karstedt catalyst was that it is more dynamic than the Speier's catalyst for heterogeneous Hydrosilylation reactions because of its capacity to frame better colloidal Pt particles during the catalyst inception step. The encompassing reactions acted in hexane were seen to continue quickly to produce elastomeric network polymers rather than the sluggish arrangement of hard, boring solids in the report by Houser et al.

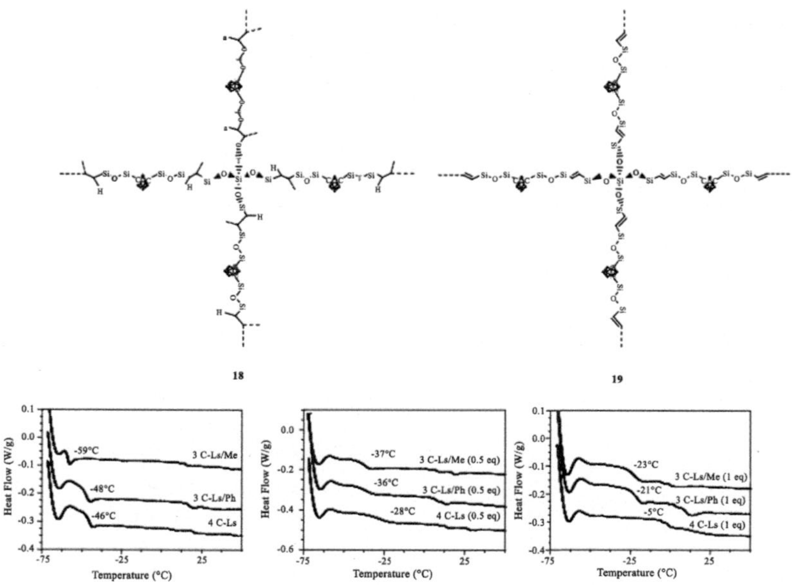

Figure 8.20. (Top) Schematic representations of the completely hydrosilated elastomeric network from 14 + 4 C–Ls (18) (left) and the partially hydrosilated

elastomeric network from 17 + 4 C–Ls (19) (right). (Bottom) DSC thermograms depicting the completely hydrosilated networks from 14 + 4 C–Ls (18) (a), and the partially and completely hydrosilated elastomeric network (19) (b) and (20) (c), respectively, from 17 + 4 C–Ls reported by Kolel-Veetil et al.

The adaptable and straightforward movies of the immersed elastomeric network polymers from the vinyl monomer had Tg values beneath −35 °C while the Tg upsides of the movies shaped from the ethynyl monomer were under 0 °C (Figure 8.20). The elastomeric polymeric organizations from 14 and 17 were found to have debasement temperatures going from 500 to 550 °C.

8.8. APPLICATIONS

8.8.1. High-Temperature and Miscellaneous

The interest in high-temperature elastomeric materials originates from the appeal for such materials for application in trends-etting innovations, especially the aviation, security, and PC ventures. Such materials are supposed to have long haul warm, thermo-oxidative and hydrolytic soundness at or more than 300 to 350 °C and to likewise can keep up with articulated adaptability to well underneath surrounding temperature.

The elastomeric crosslinked network polymers of carboranylenesiloxanes and silarylene-siloxanes portrayed in this section have comparative properties and are an ideal possibility for a wide assortment of designing applications under extreme conditions.

Likewise, these organization polymers likewise have the utility in different applications that could conceivably straightforwardly connect with their elastomeric properties and are accessible because of the presence of the presented crosslinking groups, for example, diacetylene and ethynyl. Such applications are presently being assessed.

One such application is in the thermo-oxidative bond of superior execution natural filaments. Keller has detailed the utilization of the poly(carborane-disiloxanediacetylene) polymer in the assurance of carbon filaments. Carbon strands covered with poly (carborane-siloxane-acetylene) are accounted for to shape a defensive hindrance against oxidation at raised temperatures.

At this point when utilized as a grid material (earthenware), the polymer was found to safeguard the carbon filaments from oxidative breakdown. The utilization of direct carborane-siloxane-acetylenic polymers as antecedent materials for thermosets and ceramics for composite applications somewhere in the range of 500 and 1,500 °C, separately, in an oxidizing climate was likewise portrayed. These direct polymers enjoy the benefit of being very simple to process and to be changed over into thermosets or ceramics since they are either fluids at room temperature or low softening solids. The elastomeric renditions of the poly (carborane-siloxane-acetylene) and of the hydrosilated carboranylenesiloxanes ought to be considerably more helpful for use as defensive coatings for high-performance strands as a result of their simplicity of use.

Kolel-Veetil et al. as of late depicted the insurance of superior execution natural filaments Zylon (PBO, poly (p-phenylenebenzobisoxazole), Kevlar, and carbon strands by such polymers. As these crosslinked networks are steady in air over the debasement temperatures of the natural filaments (450-700 °C), they forestall the devastating degradative oxidation of the superior exhibition strands when applied as coatings.

The diacetylene-containing carboranylenesiloxanes have been assessed to be uncommon high-temperature dielectric encasings in primer examinations. The polymers have additionally shown noteworthy high-temperature glue qualities during such examinations.

At the point when gotten in their unmistakable structure, these materials can likewise possibly work as mechanochromic sensors due to the crosslinking in the diacetylene units. The elastomeric organizations of crossover silicones ought to likewise have the option to work as high-temperature gas detachment layers.

On account of the straightforward hydrosilated organizations of carboranylenesiloxanes got by Hydrosilylation reactions, the chance exists for the utilization of the organizations in covering and sensor applications.

Furthermore, to some degree, hydrosilated network frameworks, the option exists for the reaction and detecting of particles at the natural unsaturation destinations that are consistently scattered in the hydrosilated film.On account of all of the carboranylenesiloxane network polymers delivered either by warm polymerization or by Hydrosilylation implies, the option exists for their utilization as neutron retention materials that might find applications in atomic reactor security. The presence of boron atoms in these polymeric materials that are known to be uncommon neutron retaining species by the notable boron neutron catch (BNC) reaction upgrades their significance in atomic applications.

8.8.2. Production of Ceramic Nanomaterials

The pyrolysis of different metallic subsidiaries of diacetylene-containing carboranylenesiloxanes, which has metal groups bound to the diacetylene part or being remembered for the polymer backbone, has been accounted for to bear the cost of ceramic nanomaterials with assorted directing properties. The derivatization of a diacetylene-containing carboranylenesiloxane was accomplished by the complexation of the triple bonds in the diacetylene moiety with a picked organometallic moiety. Contingent upon the proportion of the carboranylenesiloxane and organometallic moiety reactants, uncomplexed 21a, to some extent complexed 21b, and totally complexed 21c carboranylenesiloxane could be acquired in different proportions from the reaction (Figure 8.22). Before transformation into the clay, the metallic subsidiaries are changed over completely to a thermoset through the crosslinking reactions of 21a and 21b. The thermoset, in this manner shaped, was pyrolyzed to wanted temperatures to deliver the clay nanomaterials with assorted attractive and leading properties.

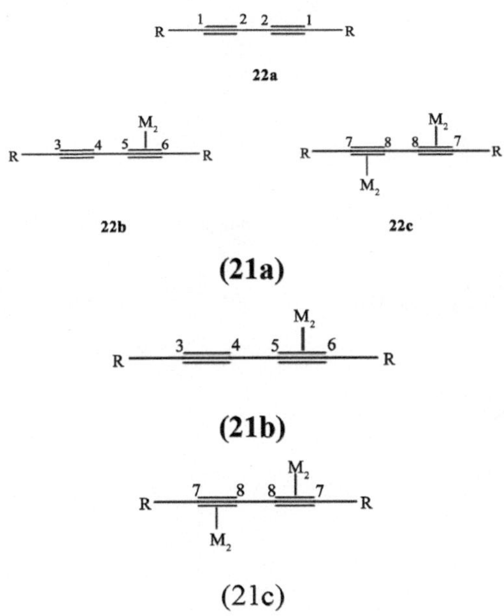

R = -SiMe$_2$-polymer M_2 = organometallic moiety

Figure 8.21. Schematic representations of the uncomplexed (21a), partially complexed (21b), and completely complexed (21c) metallic derivatives of a diacetylene-containing carboranylenesiloxane.

Utilizing this system, Kolel Veetil et al. revealed the formation of a superconducting combination of β-Mo2C and carbon nanotubes in an undefined combination of silicon and boron intensifies by the pyrolysis of the metallic item gotten from Cp2Mo2 (CO) 6 and a diacetylene-containing carboranylenesiloxane.

The equimolar reaction of Cp2Mo2(CO)6 with the carboranylenesiloxane 1 in THF was found to bring about the relocation of two of its carbonyl ligands by each triple bond in the diacetylene unit to yield a π-fortified complex. FT-IR and 13C NMR ghastly confirmations supported the arrangement of metallic subsidiaries, for example, 21b and 21c and the maintenance of some unreacted 21a. During the thermoset arrangement, the progressive misfortunes of the labile carbonyl and cyclopentadienyl

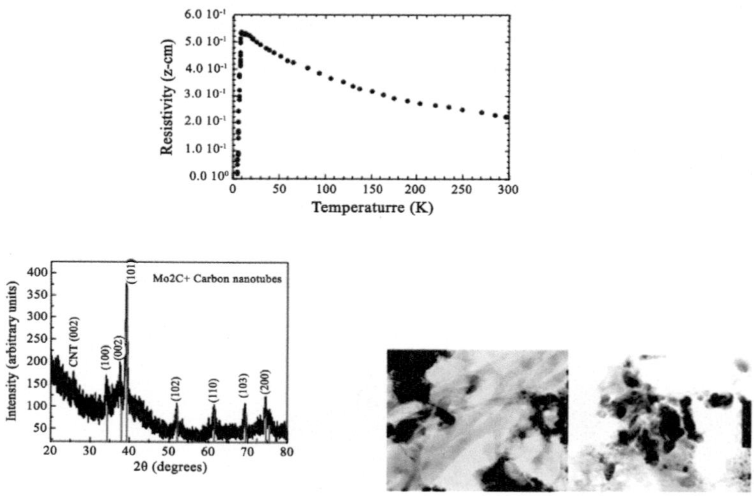

Figure 8.22. The resistivity plot, the XRD spectrum and TEM micrographs of the pyrolysis of the Cp_2 $Mo_2(CO)_4$ complex of 1.

ligands were seen from the combination around 222 and 267 °C, individually. During the pyrolysis of the thermoset, the deficiency of silicon-bound methyl groups was likewise seen around 500 °C. On finishing of the pyrolysis at 1,000 °C, accompanying arrangements of nanoparticles of β-Mo2C and carbon nanotubes were seen as confirmed by X-beam diffraction and TEM studies (Figure 8.22).

The nanoparticles were acquired in a nebulous framework of silicon and boron compounds. Resistivity investigations of the combination uncovered that it was superconducting in nature with a basic temperature (Tc) of 8 K (Figure 8.22). This addresses just the third illustration of a development of carbon nanotubes catalyzed by the metal Mo alone.

Nanomaterials of different progress metal borides, carbides and silicides with assorted attractive and directing properties have been acquired from 1 and other diacetylene-containing carboranylenesiloxanes involving a similar system as utilized in the β-Mo2C development. A portion of the nanomaterials delivered incorporate CoB, FeCoB, CuB2, and Fe5Si3.

In rundown, late advances in the space of organization polymers of carboranylenesiloxane and silarylene-siloxanes have produced remarkable high-temperature elastomeric materials because of the joining of new techniques in their synthesis. Notwithstanding their high-temperature properties, the presence of different assorted constituents in the materials has opened roads for the use of these materials in bulk applications.

8.9 CONCLUSION

The most notable properties of silicones are their exceptionally low glass transition temperatures (Tg) and low surface tension. According to Pauling, the electronegativities of silicon and oxygen are 1.8 and 3.5, respectively. Because of this variance, the Si–O bond –Si–O– has an estimated 37 to 51 percent ionic nature.

Ionic processes are used to degrade a silicone backbone containing polarizable siloxyl units. This intramolecular cycloreversion or depolymerization of silicones is thought to happen from as few as four progressive Si-O links. The pyrolysis of several metallic derivatives of diacetylene-containing carboranylenesiloxanes has been calculated to carry the expense of a large collection of ceramic nanomaterials.

The metallic subsidiary is totally converted to a thermoset before transformation into clay via the crosslinking reactions of 21a and 21b. The desirability for such materials in aviation, security, and PC businesses has sparked interest in high-temperature elastomeric materials. Such materials are expected to have long-term warm, thermo-oxidative, and hydrolytic soundness at temperatures ranging from 300 to 350 °C.

REFERENCES

1. Abd-El-Aziz, A., Carraher, C., Pittman, C. and Zeldin, M., 2008. Inorganic and Organometallic Macromolecules. [online] Available at: <https://link.springer.com/book/10.1007/978-0-387-72947-3> [Accessed 30 June 2022].
2. Homrighausen, C. and Keller, T., 2001. High-temperature elastomers from silarylene-siloxane-diacetylene linear polymers. *Journal of Polymer Science Part A: Polymer Chemistry*, [online] 40(1), pp.88-94. Available at: <https://onlinelibrary.wiley.com/doi/10.1002/pola.10091> [Accessed 30 June 2022].
3. Koh, K. and Sohn, H., 2021. Fast Curable Polysiloxane-Silphenylene Hybrimer with High Transparency and Refractive Index for Optical Applications. *Polymers*, [online] 13(4), p.515. Available at: <https://www.ncbi.nlm.nih.gov/pmc/articles/PMC7915534/> [Accessed 30 June 2022].
4. Koide, N. and Lenz, R., 2007. Preparation and properties of poly(silarylene siloxane)s. *Journal of Polymer Science: Polymer Symposia*, [online] 70(1), pp.91-105. Available at: <https://onlinelibrary.wiley.com/doi/10.1002/polc.5070700108> [Accessed 30 June 2022].
5. Stewart, D., Peters, E., Beard, C., Dunks, G., Hedaya, E., Kwiatkowski, G., Moffitt, R. and Bohan, J., 1979. D₂-m-Carborane Siloxanes. 7. Synthesis and Properties of Ultra-High Molecular Weight Polymer. *Macromolecules*, [online] 12(3), pp.373-377. Available at: <https://pubs.acs.org/doi/10.1021/ma60069a004> [Accessed 30 June 2022].

INDEX

C

N

O

T

U

V

W

X

Z